Water Utility Management

AWWA MANUAL M5

Second Edition

American Water Works Association

Science and Technology

AWWA unites the entire water community by developing and distributing authoritative scientific and technological knowledge. Through its members, AWWA develops industry standards for products and processes that advance public health and safety. AWWA also provides quality improvement programs for water and wastewater utilities.

MANUAL OF WATER SUPPLY PRACTICES—M5, Second Edition

Water Utility Management

Project Manager: Neal Hyde
Production Editor: Neal Hyde

Library of Congress Cataloging-in-Publication Data

American Water Works Association.
    Water utility management.—2nd ed.
        p. cm. — (AWWA manual ; M5)
    Includes index.
    ISBN 1-58321-361-9
      1. Waterworks–Management. I. Title. II. Series.

TD491.A49 no. M5
[HD4456]
628.1 s–dc22
[363.6'1'068]

2005048284

Printed in the United States of America

American Water Works Association
6666 West Quincy Avenue
Denver, CO 80235-3089

ISBN 1-58321-361-9

Printed on recycled paper

Contents

This page intentionally blank.

Figures

This page intentionally blank.

Tables

This page intentionally blank.

Acknowledgments

This manual was authored by Kay Kutchins, of Kay Kutchins and Associates Inc., San Antonio, Texas. Her experience includes work in utility organization structures and behavior, administration, human resource management, and internal and external communication. She is an Honorary Member, past national Vice President, and council member of the American Water Works Association. She is also a member of the Association for Women in Communication, the Society for Human Resource Management, a Fuller Award recipient, and recipient of the AWWA Outstanding Service to the Water Industry recognition.

The author gives special thanks and recognition to the following individuals whose contributions aided in the development of this manual:

Carolyn Ahrens-Weiland, Booth Ahrens Werkenthen, LLP, Austin, Texas
Charles F. Anderson, City of Arlington, Arlington, Texas
J. Paul Blake, Seattle Public Utilities, Seattle, Wash.
Michael K. Dutton, Kay Kutchins & Associates Inc., San Antonio, Texas
Ross A. Flaherty, President, Flaherty & Associates, Inc., San Antonio, Texas
Richard Gerstberger, TAP Resources Development Group Inc., Denver, Colo.
Rhonda Harris, Professional Operations Inc., Dallas, Texas
Ronny Hyde, Fort Worth, Texas
J. Rowe McKinley, Black & Veatch, Overland Park, Kan.
*Sharon Metz, City of Columbus, Columbus, Ohio
Wendy Nero, CH2M Hill, Deerfield Beach, Fla.
Alison Posinski, City of Cincinnati, Cincinnati, Ohio
Harold Smith, Raftelis Financial Consulting, Charlotte, N.C.
William G. Stannard, Raftelis Financial Consulting, Kansas City, Mo.

AWWA's Management Division Trustees and Committee Chairs provided invaluable assistance in reviewing the manual and in suggesting content.

George Raftelis, Division Chair, Raftelis Financial Consulting, Charlotte, N.C.
Randolph Brown, City of Pompano Beach, Pompano Beach, Fla.
Mike Gritzuk, City of Phoenix, Phoenix, Ariz.
Dale L. Jutila, City of Gresham, Gresham, Ore.
Alison Posinski, City of Cincinnati, Cincinnati, Ohio
Arlene Post, Las Virgenes Municipal Water District, Calabasas, Calif.
Karen Snyder, Snyder Communications, Hendersonville, Tenn.
Marsi A. Steirer, City of San Diego, San Diego, Calif.
Andy Kieser and Lou Eddy, Management Development Committee
*Jim Kelly, Past Chair, Finance, Accounting and Management Controls
Harold Smith, Competitive Practices Committee, Raftelis Financial
     Consulting, Charlotte, N.C.
Jill Schwarzwalder, Health and Safety & Environment Committee
Tom Linville, Security Committee, Contra Costa Water District, Concord, Calif.

---

*Retired

J. Rowe McKinley, Rates and Charges, Black & Veatch, Overland Park, Kan.
Don Schlenger, Customer Service Committee, Cognyst Consulting LLC,
    Kinnelon, N.J.

This edition was reviewed and approved by the AWWA Management Division Board
of Trustees and the AWWA Technical & Educational Council.

The Management Division had the following personnel at the time of approval:

George A. Raftelis, Chairman, Raftelis Financial Consulting, Charlotte, N.C.
Marsi A. Steirer, Vice Chair, City of San Diego Water Dept, San Diego, Calif.
Dale L. Jutila, City of Gresham, Gresham, Ore.
Karen P. Snyder, Snyder Communications, Hendersonville, Tenn.
A. Randolph Brown, City of Pompano Beach, Pompano Beach, Fla.
Stephen J. Densberger, Pennichuck Water Service Corporation, Merrimack, N.H.
Alison F. Posinski, Greater Cincinnati Water Works, Cincinnati, Ohio
Arlene E. Post, Las Virgenes Municipal Water District, Calabasas, Calif.
Rick Harmon, Staff Advisor, AWWA, Denver, Colo.
Lois Sherry, Staff Secretary, AWWA, Denver, Colo.

The author wishes to acknowledge AWWA staff members Beth Behner, manual
coordinator, Neal Hyde, technical editor; and Lois Sherry, management division staff
secretary, who provided guidance, detailed information, encouragement, and motivation through the writing and editing processes; also, Linda Reekie, who, with the
Awwa Research Foundation staff, provided linkages to research projects that explore
management issues.

Chapter 1

Introduction

Bookstores contain rows of academic treatises, quick-study manuals, and discourses on management, motivation, leadership, and a host of other topics. Their shelves are full of books on leadership and management that may not fit the water utility industry. Various authors write that management is either an art or a science or both, and some infer that it is neither.

The American Water Works Association (AWWA) publishes Manual M5, *Water Utility Management*, because utility managers are in a challenging profession, and managerial leadership is key to successful utility operations. Utility managers guide organizations that fulfill a need essential to the public health of the communities they serve. Utility managers are responsible for environmental quality and quality-of-life issues in their communities. These managers face continuing demands to provide excellent service to customers, and what they provide is compared with the types of customer services offered by nonutility businesses and industries. Utility managers are expected to perform their tasks within budgets based on generated revenues that are difficult to increase. Utility managers are asked to ensure the lifeblood of communities that often do not know or understand the value of the water they receive.

AWWA is uniquely positioned to share managerial experience and knowledge from and among its members. This manual provides a ready reference for water utility management and leadership. The manual contains information to help improve the managerial skills described in Table 1-1.

Throughout this manual, several terms are repeated:

- vision—a clear understanding by managers of where a utility is going and what it should be

- mission—what the utility is and what its purpose is

- objectives—the goals set to maintain the mission and achieve the vision

- goals—the specific responsibilities of individuals and work teams that contribute to achieving one or more objectives

1

Table 1-1   Skill areas

Managing the Utility
- The management process
- Governance and governing board relationships
- Stewardship/ethics
- Organization structure and change management
- Building utility leadership
- Regionalization

Communications
- Recognizing publics
- Routine communications
- Internal communications
- External communications
- Public involvement
- Effective media relations
- Crisis communications

Customer Service
- Fundamental principles
- Key customer service functions
- Customer service policies
- Consumer confidence reports
- Customer interaction
- Continuous improvement

Financial Management
- Budgeting
- Rate design
- Related charges
- Capital financing
- Accounting practices and management controls
- Funding alternatives

Environmental Health and Safety
- Utility management responsibility for safety
- Environmental legislation
- Risk management and insurance
- Public notification protocols
- Occupational health and safety

Engineering Considerations
- Planning
- Design
- Construction
- Standards
- Managing contracts
- Training

Operations and Maintenance
- Record keeping
- Mapping
- Safety
- Training
- Preventive maintenance
- Scheduling
- Planning
- Human resources
- Technology
- Communication
- Equipment and tools
- Policies and procedures
- Outsourcing
- Specifications and standards
- Professional memberships
- Support services
- Regulatory matters

Information Systems and Services
- Key elements of information technology
- Information technology and information systems infrastructure
- Web technology
- Managing data
- Supporting information technology
- Standards
- Implementing new systems
- Emerging issues

Legal and Regulatory Planning
- Selecting appropriate counsel
- Managing the attorney–client relationship
- Specialized matters

Human Resources
- Staffing the utility
- Training and development
- Performance management
- Employee relations
- Disciplinary action
- Policies and procedures
- Succession planning and professional development

Security and Emergency Planning
- General security issues
- Vulnerability assessments
- Emergency response plans

- actions—the specific steps taken or processes that contribute to successful achievement of goals and objectives

- effectiveness—doing the right thing at the right time

- efficiency—running utility operations in the most cost-efficient, highest-producing manner

Water utility managers and leaders are challenged to learn how to effectively overcome organizational resistance to change. It is no secret that all human beings are limited in their ability to change, with some more limited than others. Successful managers embrace change and create a climate in which change is recognized—and perhaps welcomed—as a normal way of conducting business. By studying this manual, water utility managers learn how to cope with, and encourage, change within their utility.

This manual is different, compared to other management books, because it was developed through a collaborative effort of managers and leaders in the utility profession who successfully met the challenges of managing a water utility. They demonstrated through time and peer review the management and leadership ability that made their organizations examples of best utility practices.

Managers of municipal, nonprofit, and investor-owned water systems face competitive challenges. Each type of utility has its own governance system, elected or appointed boards of directors, or political or profit foundations that determine how decisions are made, resources utilized, and policies formulated. However, utility managers and leaders recognize a narrowing gap between nonprofit and for-profit organizations, caused by the entry of global for-profit organizations into the more traditional "nonprofit" arena through operating contracts, limited outsourcing, and outright acquisition. A well-informed utility manager is more able to effectively lead his or her utility to new achievements in a fast-paced, highly regulated, and technologically complex business environment.

Being a manager and a leader is a challenging opportunity. Being a manager implies the use of a process—including planning, organizing, directing, and controlling—and communicating that process and its components to utility staff. Being a leader requires a commitment to an organizational culture that encourages respect and motivates all employees to adopt common goals and objectives and to adapt personal interests to the improvement of the organization and its customers.

Figure 1-1   Management is a challenging opportunity

Managers have several roles within their organizations. The roles include

- interpersonal roles, such as leadership, liaison, and figurehead
- informational roles, such as monitor, disseminator, and spokesperson
- decision-maker roles, such as troubleshooter, entrepreneur, resource allocator, and negotiator

Just as managers have several roles within their organizations, managers also share several common responsibilities. These include

- staff development
- organization effectiveness
- organization relationships
- organization morale
- interpersonal relationships
- personal productivity
- self-development
- proactive thinking
- creativity
- assistance to higher-level and subordinate managers

Managers are expected to develop and implement policies and procedures adopted by the organization's governance structure and to review and approve organizational policies and procedures developed to ensure consistency in operations, maintenance, administration, finance, etc. Managers often have an opportunity to provide input into those policies and should always provide feedback concerning policy implementation, especially if implementation leads to problems or inconsistencies with good management practices.

One of a leader's greatest challenges is time management. There is never enough time to accomplish all that needs to be done. There must be time for thinking, for evaluating, for listening to others, and for communicating. Managers must learn the fine art of delegating—providing employees with the authority and the responsibility for action and holding them accountable for results. Effective delegation is a difficult task that managers must master, but mastering it gives better control of their time.

Effective managers also develop ways to organize their efforts so as to limit the amount of stress incurred. Balancing the demands of utility management with the growing importance of nonwork commitments is a challenge for the best manager. Physical well-being, adequate exercise, having a variety of interests, and proper diet all contribute positively to the ability to juggle the tasks and responsibilities that accompany utility management.

In addition, water utility managers must anticipate the changes that will occur during their managerial career and identify emerging issues that will impact the resources available. Effective utility managers should consider these questions:

1. What is the appropriate response to the increased level of competition faced by the utility and the demand for higher productivity within the organization?

2. What is the best way to manage increasingly rapid change?

3. What are the impacts of global events on operational decisions?

4. How does a diverse population in the utility's service area impact service delivery and employment issues?

5. How does economic volatility affect rates, fees, payments, collections, and the ability to deliver services?

6. How do demographic shifts affect service delivery, infrastructure construction and replacement, budget, and capital planning?

7. What is the best way to expand the knowledge, skills, and abilities of utility staff, especially if funding for investment in human capital shrinks?

8. How can the supply of quality applicants and employees in the workforce be improved?

9. How do increasing interest in quality-of-work issues, increased emphasis on health and wellness, and the changing nature of family relationships and families affect the work environment?

10. What is the best way to effectively address the increased attention to the role of business and government in society, including ethics, responsibility, and fiscal conservatism?

11. What is the appropriate way to deal with the challenges of an aging population, including work assignments, technological training, benefits costs, and a much older retirement age than the historical age of 65?

12. What is the best way to manage small, self-contained work groups and a decentralization of work in an era when staff may have flexible, nontraditional working hours?

An individual manager's knowledge, skills, abilities, vision, intuition, and other attributes govern how the manager performs. This manual offers ideas about successful management. The steps each individual takes to try something new, to implement best practices, and to grow personally and as a utility manager determine success. The objective is to make a difference as a water utility manager and professional.

This page intentionally blank.

Chapter **2**

Managing the Utility

OVERVIEW

Utility managers must know and administer many diverse elements in their daily activities. That process begins with a vision of what managers expect their utility to be. They must fully understand the legal and regulatory framework under which they operate, as well as the utility's mission, and they must adapt to rapidly changing legal, regulatory, and customer demands. With the general trend being toward constrained resources and increased expectations, managers must continually analyze costs; reallocate available funds, staff, and equipment; and, maintain a highly motivated and productive staff. The need for improved interpersonal, organizational, and external communications has never been greater. And, the demands of stakeholders for accountability in the acquisition and use of resources requires greater attention to documentation of work processes and control of costs and program expenditures.

Powerful trends are shaping the water profession and will continue to impact what has been, for the most part, a traditional business model with few entrepreneurial characteristics and low tolerance for risk. These trends include

- widespread competition
- increasing regulations
- scientific uncertainty
- political intervention
- increasing public expectations

In many respects, management is a process that promotes continuous improvement of a utility's business practices. The management process should be thought of as an umbrella that is used to both shield managers from the elements and allow them to identify and maneuver through their work environment. Example of elements include

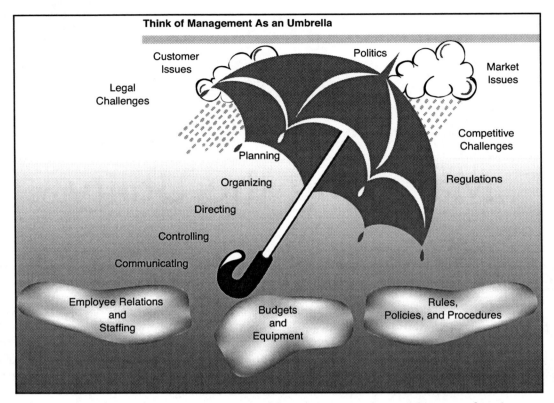

Figure 2-1   The management process can shield managers and provide more freedom

- many externalities that are addressed on a daily basis—legal challenges, customer expectations and issues, politics, regulations, competitive challenges, and economic challenges in a market economy

- work environment includes the internal constraints all managers face—staffing issues, employee relations, budgets, equipment needs and uses, rules, and policies and procedures

- culture and communications include the style of the organization and the embedded values that direct all of its functions and activities

- management process includes the planning, organizing, directing, and controlling processes essential to efficient utility operations

A utility manager must communicate the utility's vision and mission to both external stakeholders and employees; it may be the primary duty. It is through this communication linkage that support is built for utility programs and projects. The manager must know and be able to translate long-term utility planning processes into terms that external stakeholders understand and can support. Managers must also listen to and act responsibly on stakeholders' concerns that impact the planning process. Managers must marshal and allocate resources to those priorities that address both utility and stakeholder needs. And, they must delegate authority and hold subordinates accountable for the majority of day-to-day operating decisions, coordination of efforts, and successful achievements of the utility's strategic objectives.

THE MANAGEMENT PROCESS

Planning

A hallmark of a successful utility is its incorporation of planning for its future rather than reacting to events as they happen. Every activity a manager engages in merits the time it takes to plan the activity, anticipate responses from others, and integrate new information into the activity. There are several types of planning that should be considered, including operational, administrative, project, financial, organizational, communications, and emergency:

- operational planning includes reactive and preventive maintenance requirements, as well as equipment and vehicle maintenance. it also includes acquisition and allocation of resources.

- administrative planning includes staffing, information acquisition and analysis, materials management, and performance management.

- project planning must consider scheduling, contracting, public involvement and outreach, and oversight/program management.

- financial planning must consider revenues, expenses, reserve requirements, access to capital, and surplus and deficit issues.

- organizational planning considers functional alignments, staffing and succession issues, business processes, and competitive issues.

- communications planning includes internal and external communications strategies, messages, programs, community outreach efforts, and media relations.

- emergency planning includes vulnerability to system disruptions (whether through natural or human occurrences), health and safety issues, and consideration of the impact of emergencies in other parts of the community on the utility's ability to meet its mission.

Specific references to planning processes in each of the above areas appear elsewhere in this manual. In addition, AWWA offers a series of manuals relating to more than 50 aspects of water utilities that address the topics in greater detail. Information can be obtained through the Association's customer service center.

An effective planning tool is the development of a task list associated with the proposed program or project. Such a task list, prepared with broad input from all parts of the utility and those external stakeholders who are directly or indirectly involved, achieves several objectives. First, it employs a brainstorming process that enables all stakeholders to identify their areas of interest—location of project, community concerns, environmental issues, financial concerns, scheduling, etc. Second, it gives all parties an opportunity to look at and set priorities for the project, resulting in a well-defined strategy before the project begins. Third, this level of planning can effectively reduce the costs associated with the project, including administrative review, change orders, meeting requirements, public response, and acquisition of materials/supplies/equipment on a *just-in-time* basis.

Some Important Planning Reminders

- **The governance body** sets overall goals and objectives, works with staff to set action plans, completes financial planning, and approves capital planning.
- **Make time for planning.** It does not just happen.
- **Encourage participation.** Those who are affected by goals, objectives, and action plans should participate in their creation.

Effective planning benefits utility managers in many ways. It makes budget justification easier and helps set priorities for programs and task activities. It motivates both staff and governing bodies by establishing a vision of the future. It encourages clear coordination and accountability for actions and helps avoid mistakes, oversights, and vague responsibilities. It assures sufficient lead time for projects and programs and defines clear measures for success. Moreover, it leads to steady growth and rational implementation of changes by encouraging annual reevaluation of the utility's vision, mission, goals, and objectives.

Organizing

Organizing is that part of the management process that uses information generated during the planning process to make decisions about obtaining and allocating resources, structuring work processes, and managing project or program components. It involves knowledge of staff capacity (especially multiskill capabilities), and of easily accessed internal and external resources available for the duration of the project or program. It also requires the manager to presume that planning information is incomplete and that components of the planning process are subject to reconfirmation to ensure that decisions have the highest level of confidence of success.

The specific nature of the project or program determines whether traditional work processes are appropriate or whether modifications are necessary. Areas for consideration include work team composition, equipment allocation, safety and security provisions, information availability, and access to specific resources either within or external to the utility. It also dictates the management strategies for each project or program component.

Directing

Directing is the part of the management process in which the manager delegates responsibility and authority for the project or program, defines the accountability (expectations) of the functional unit, and supervises, in one or more ways, the achievement of the assignment.

Delegation is a term familiar to most managers, but delegation is frequently not as effective as desired. Delegation may fail if a manager lacks confidence in subordinates or coworkers or if a project or program is delegated to an individual who lacks experience or expertise in the area. Staff development addresses this issue by allowing individuals to develop skills and gain the confidence of superiors. Time constraints also impact delegation, especially if a project or program needs immediate attention with little or no time for review. Also negatively affecting the success of delegation is inadequate delegation—giving a subordinate or coworker responsibility for a program or project but not the authority to make decisions about resources and tasks.

Supervising is the part of the manager's responsibility that recognizes the work process must be completed by others. It includes providing direction and guidance concerning the project or program, teaching and demonstrating portions that are unfamiliar to the work team, and monitoring the effort—from inception through completion—to ensure that it is completed according to plan and within budget and time requirements.

Controlling

Controlling is the part of the management process in which managers make use of project or program status information during and after the project to identify ways in which

- human and material resources can be more effectively used

- costs can be controlled

- time schedules can be more realistic

- information gained can be analyzed and incorporated into planning future projects or programs

- ongoing assessment of information can improve business processes and resource allocation

GOVERNANCE AND GOVERNING-BOARD RELATIONSHIPS

A utility's governance is contained in the legal framework under which it operates. It may be

- a municipal water utility, constituted as a department of a city, county, or provincial government

- a municipal water utility chartered independently of municipal rule

- a public water system chartered as a special-purpose district

- a rural water system

- an investor-owned water system

- another, authorized organizational structure

In all instances, a governing body is constituted either by appointment or by election and is responsible for establishing the vision and mission of the utility; supervising finances to ensure self-sufficiency and sustainable development; appointing, directing, and evaluating the utility's management; establishing and approving general policies; authorizing annual budgets, major expenditures, contracts, and issuance of debt; and approving the utility's master plan and related capital plans.

The utility is directed by a chief executive officer or general manager who reports to the governing body and carries out its decisions. This senior executive officer is accountable to the governing body for the ongoing administrative, financial, technical, and operating activities of the utility and may be directly involved in managing some of those functions, depending on the size of the utility and extent of its staff.

Figure 2-2   The single most important job requirement for a utility manager may be the management of his or her governing body

Utility policy makers—whether they are appointed or elected boards of directors or elected public officials (councils, commissions, etc.)—have nine principal responsibilities. Utility managers need to fully understand those responsibilities in order to assist their governing bodies with necessary information and staff support for decision making. The following describes these responsibilities:

1. **They ensure that the utility achieves its stated purpose and objectives.** They provide community or owner input into and approval of the overall vision of the utility and its critical missions and approve annual action objectives, capital and financial plans, and special programs.

2. **They establish policies and procedures that govern the utility's operations.** Policies and procedures should govern customer services, financial and administrative matters, personnel rules and regulations, adopted standards for professional services and system development, regulatory compliance, ethics and conflict of interest, and a broad range of utility-specific matters. In conjunction with this, they

   • provide management with a scope of authority and standards of responsibility and performance accountability. this includes a requirement that management report back to the governing body before taking action that exceeds that authority
   • establish committees to carry on the governing body's function of policy development and define rules as to how those committees work among themselves and with staff
   • require that written policy documents be approved as part of the governing process, with approval conveyed through meeting minutes and policy documents entered into the utility's official records

3. **They raise and manage the utility's funds.** This includes reviewing and developing the utility budget, master plan, and capital improvement programs and approving the final documents. It also includes evaluating the needs for increases in water and sewer rates and charges, evaluating the need for special fees, issuing debt for capital projects, and providing oversight of control systems that manage corporate funds. Finally, they authorize the annual audit of the utility according to generally accepted

accounting practices, receive and approve the audit and accompanying management letter, and assign and direct any business corrections recommended by the auditor.

4. **They employ the utility's senior executive; set personnel policies, wages, and benefits; and may act in the final appeals step in adverse personnel actions.** They ensure that position descriptions are complete and reflect current requirements, ensure compliance with all fair-employment laws and regulations, approve and adopt personnel policies, and adopt compensation and benefit plans that are fiscally responsible and attractive to both current employees and potential applicants for employment. They hold managers accountable for hiring those staff members who directly report to them and for all human resource processes throughout the organization.

5. **They supervise and evaluate the performance of the utility manager.** They regularly review performance expectations, including, but not limited to, operational reports, financial reports, productivity measures, customer complaints and utility responses, capital program progress, etc.

6. **They appoint committees of the board.** Committees run the range of governing-body involvement with the utility, but usually include long-range planning, finance, personnel, legislative, policy, and community relations. Governance of the committee and regular reporting formats may be established by the governing body.

7. **They hold property for the benefit of the utility, as appropriate to state or provincial law.** Applicability, and the items covered, are unique to each governing body, but must be clearly spelled out in governing documents.

8. **They develop successors for governing positions and ensure leadership on behalf of the utility.** They insist on a broad knowledge of utility operations and issues and on direct, open, and honest communications.

9. **They employ consultants and specialists who report to the governing body.** These normally include the independent auditor (either internal to the organization or external for financial audits), general and special counsel, and legislative liaisons. They may also include a consumer advocate, a mediation specialist, security specialists, and others.

Utility managers should provide new governing body members with a comprehensive orientation to the business practices, physical locations, and culture of the organization as early as possible in their tenure. Continuing communication and work with governing committees enables the utility to maintain its message to, and credibility with, its governing agency. Indeed, the most important job requirement for a utility manager may be the management of his or her governing body. At the same time, the manager and all staff members must be aware of and compliant with state or provincial laws regulating political activity so that there are no opportunities for inappropriate or illegal involvement with governing body members.

Key Strategies for Governing-Body Effectiveness

1. Have an informed governing body; start new members off right.

   - inform each member about the current vision, mission, goals, and objectives.
   - provide each member with bylaws or governing documents, a written description of the utility, and a current organization chart.
   - provide special or key reports from the prior year, the most recent annual report and annual audit, and financial reports (budgets and actual expenditures) for the past year.
   - list committee charges and assignments, and provide minutes of meetings held by committees or the governing body for the prior six months.
   - provide a roster of board members and key staff.

2. Train the governing body.

   - provide site visits with briefings from senior staff on programs and services and introduction to staff.
   - provide an overview of administration, organization structure, governing body and staff roles and responsibilities, and a copy of the utility's adopted governing body manual, including policies and procedures.
   - review financial issues, including budget, investments, property, rates and charges, and financial policies.
   - provide committee orientation, including appropriate documents.

3. Stress the importance of reaching consensus and acting for the greater good of the utility and the community.

   - encourage participation in discussion of all assumptions and issues.
   - obtain all data relating to issues and consider all alternatives before making decisions.
   - determine the extent to which public involvement is part of the decision process and have a clear method of managing that involvement so decisions are made.

4. Set and practice ethical standards, and demand ethical behavior by all utility staff in daily business operations.

   - know and abide by conflict-of-interest laws.
   - establish and abide by an ethics policy.

5. Maintain records of governing body actions and documents.

   - establish and post agendas for meetings.
   - maintain minutes of meetings (electronic and formal documents).
   - conform motions to agenda wording.
   - maintain official records, files, actions, contracts, and legal documents and retain documents according to state or provincial law.

STEWARDSHIP AND ETHICS

Utility governing bodies, managers, and employees are in the public eye, regardless of whether the utility is investor-owned or a public entity. They are responsible for effective stewardship of the resources to accomplish their required mission and objectives. This includes staff, financial resources, procurement, equipment and vehicles, materials, and other things of value. Responsible stewardship includes but is not limited to

- clear personnel policies concerning work requirements, hours, duty stations, fair employment standards, and disciplinary practices

- clear financial policies concerning purchase of items from a master purchase order, petty cash, per diem, and expense reimbursement

- procurement policies that provide for transparency in competitive bidding, established protocols for obtaining professional services, and compliance with state or provincial procurement laws

- clear policies concerning the use of vehicles and equipment, both during work and nonwork hours. other requirements include compliance with tax requirements for employer-provided vehicles, maintenance of records for gasoline usage, use of utility computers and telephone equipment, and provisions for adverse actions when theft of utility property or service is found

- procedures by which employees may acquire surplus equipment, vehicles, materials, or supplies no longer needed or used by the utility (cars, trucks, computers, pipe, hand tools, etc.)

In the wake of media attention to corporate and public malfeasance, there is an escalating challenge for utility managers and governing bodies to take strong action concerning ethics and conflicts of interest. Many communities, often at the behest of citizen panels, have enacted strong ethics ordinances to enforce ethical behavior and disclosure of conflicts of interest that might compromise decisions made by staff or by a governing body. Most states and provinces have conflict-of-interest or ethics statutes that cover the public policy process for themselves and, by extension, municipalities and other publicly chartered agencies. Investor-owned utilities also maintain ethics and conflict-of-interest policies and procedures.

Such policies generally require employees and policy makers to, during working and nonworking hours, act in a manner that inspires public trust in their integrity, impartiality, and devotion to the interests of the utility and its ratepayers. Policies generally specify that there is no offering, acceptance, or solicitation of any money, property, service, or other item of value by way of gift, favor, inducement, or loan offered or received with the intent that there is any influence by such conduct in connection with an employee's or official's discharge of public duties. Finally, policies generally require that each employee or official disclose such conflicts that tend to or could result in the erosion or destruction of public confidence in the operations of the utility. General prohibitions under such policies include

- engaging in any business or professional activity that would impair their independent judgment or discharge of official duties

- holding a direct or indirect financial interest in any business entity, business transaction, or business endeavor that creates or appears to create a conflict of interest

- disclosing any transaction or business activity that is or appears to be a conflict of interest

- ensuring documentation of disclosures

- disclosing confidential information concerning property, operations, policies, or personnel activities of the utility

- disclosing any personnel actions, unless discussed in open meetings or otherwise officially disseminated

- using official positions or facilities to secure special advantages

- receiving any gift of money or item of value from an organization that is regulated by, contracts with, or anticipates contracting with the utility. in most cases, provision is made for a particular financial value under which gifts may be accepted

- transporting unauthorized persons in utility vehicles unless properly authorized to do so

Employees whose work provides them with access to confidential information about personnel, proprietary practices, or utility activities must be aware that they are not to disclose such information without the written authorization of senior management or legal counsel.

ORGANIZATION STRUCTURE AND CHANGE MANAGEMENT

Water utility organizations experience the same types of competitive pressures that characterize the private sector business model. Most utility organization structures are built as a vertical entity with multiple levels of relatively small spans of supervision. As a business practice, many organizations—utility and nonutility—are transforming themselves into horizontal organizations with broader spans of control and fewer levels of supervision. This transformation brings decision making to lower levels of the utility and supports mandates to contain costs and maintain services. While the water industry may not see the type of deregulation that has characterized other utility industries, it is—and will continue to be—in the midst of a radical change in the business environment. Cost containment, outsourcing, expanded automation and information management, increased politicization of utility planning, changing customer expectations, and changing customer attitudes confront utility managers at every corner.

Traditional issues and pressures are still there. Regulatory requirements are expanding, infrastructure is deteriorating, capital requirements are increasing, ratepayer demands for improved quality and reduced costs are increasing, customer service and billing standards are being driven up by the level of service provided by other industries, and institutional distrust issues continue to spill over into utility operations.

Compounding those pressures are the multitude of organizational and managerial problems, such as organizational inefficiencies, employee low trust of management, communication breakdowns, organizational silos, low personal accountability, inefficient work process, excessive policies/procedures/rules, and internal disruption. To maintain effectiveness, utility managers must understand the organization itself and exercise a new level of leadership skill.

Knowing the Organization and Its Mission

Effective managers must know their organization in order to make appropriate and essential decisions. A critical skill is the ability to identify and focus on core organization issues rather than on perceived symptoms of issues confronting the organization. Managers must know the utility's mission or must lead the utility organization in developing a vision of what it wants to be and in defining its mission. The strategic decisions and actions that follow add value to the utility's efforts in serving its customers.

A utility's vision, mission, and strategic objectives must be clearly understood throughout the organization, and successful managers must continually communicate the utility's business message to both employees and customers.

Strategic/Tactical Planning, Implementation, and Management

Business and organizational theorists have engaged in widespread research about organization behavior. As business globalization continues and as competitiveness requirements expand, organization research will continue. Strategic and tactical planning focuses on four areas:

- operational excellence—the utility delivers reliable and dependable products and services at competitive prices with minimal disruption or inconvenience

- customer-centered—the utility's focus is on the customer, and the utility maintains a thorough knowledge of its customers, customer groups, and the services they need

- product/service leadership—the utility leadership "thinks outside the box" to offer products and services that expand its current boundaries

- enrichment strategies—the utility helps customers and employees fulfill their potential and reach higher levels of development

Confusion and organizational mistrust are the minimal outcomes if management initiatives are not clearly linked to utility strategies. It becomes more likely that the utility will veer off course in its efforts to accomplish its mission. Once employees are clear about the mission or strategy, managers can look more intensively at core organization culture and leadership practices.

The Utility's Organizational Culture

Every organization has its own culture—the way it behaves and the way it has always done things. That culture drives how business processes are implemented and how success is achieved and measured. Successful managers understand that for ideas and processes to work they must fit in with the culture and values of the organization. Culture describes the way of life for the organization by

- providing consistency, order, and structure
- establishing an internal way of life
- determining the conditions for internal effectiveness

Culture is also a major contributor in setting expectations and priorities, determining the nature and use of power, determining whether or not teamwork is important and expected, and providing the framework for addressing, managing, and resolving conflicts. In essence, culture determines how a utility plans its work, organizes and coordinates activity, builds teams, manages and assesses performance, and gets the results deemed important.

While each organization may exhibit different cultural characteristics, these characteristics generally fall into four categories: collaboration, control, competence, and cultivation. An organization may share traits of some or all of these, but one of the four is usually dominant.

- *Collaboration cultures* value synergy, are dedicated to customer satisfaction, and exhibit teamwork, close coordination, and cooperation. Collaboration culture managers tend to exhibit participatory styles and expect significant involvement from all organizational levels.

- *Control cultures* value certainty and control of the factors of success in goal achievement, including, but not limited to, predictability, safety, accuracy, and dependability. Control culture managers are often directive in their leadership style.

- *Competence cultures* value distinction and conceptual goals and seek to offer unparalleled and unmatched products or services. Competence culture managers are often visionaries, take long-range views, develop appropriate strategies, and try to anticipate any contingency.

- *Cultivation cultures* value enrichment and seek to ensure the highest level of customer satisfaction and support while connecting the utility's values and ideals with operational actions. Cultivation culture managers have great faith and optimism in the integrity and untapped talents of their people and often have charismatic personalities.

The cornerstone of organizational effectiveness is customer value. It is essential that everyone in the organization—not just the leadership team—clearly communicates how the utility provides significant added value to its customers. Until this element becomes ingrained in the organization's culture, other change activities are compromised. Thus, managers must ask themselves three basic questions:

- Is the organization clear about its core strategy?

- Is that core strategy truly functioning as a guiding principle for the organization?

- What role do other support strategies play in supporting the core strategy?

Managers must place a high priority on understanding and confirming organization culture before they consider or attempt any organizational change program. It is not just another variable—it is a fundamental principle. Management ideas succeed or fail in large part based on how well managers recognize and honor the power of organizational culture. Successful managers manage in a way that complements the organization's core culture.

BUILDING UTILITY LEADERSHIP

Utility managers are expected and required to be good leaders who establish a climate that unites the organization's employees in a commitment to making it successful. Thus, leadership becomes an essential element in progressive approaches to managing and leading utility organizations and is a primary skill. Leaders come in all shapes and sizes, are found in all parts of the organization, and use a variety of styles, communication strategies, and knowledge/skill/ability sets to authenticate their leadership position. Most of all, they project a demeanor that results in recognition by others that they are worthy of being a leader, with all the respect and collegiality that accompanies the title.

Utility leadership is the capacity or ability of a person or persons to envision the future needs, structures, and behavior of the organization and its stakeholders and to influence the actions of others to achieve that vision and its attendant mission and goals. Traditionally, utility leaders have focused on the systems and institutions that have done things well for stakeholders. That focus includes administration, initiation of measured changes, and long-term vision of and related planning for technical aspects of utilities, but relatively short-term vision as to institutional and stakeholder expectations. In this model, leaders have asked *how* and *when*.

The emerging utility leader is an innovator with a planning horizon that begins with a long-term definition of *What do we want to be?* and progresses to develop and empower people who achieve that objective. In this model, leaders ask *what* and *why* and frequently challenge the status quo. They create and espouse confidence in the ability of the organization and its resources to achieve the vision, but continually refine or redefine that vision to meet changing conditions. Leaders create the work climate that draws all utility personnel into committing to the organization's success. They encourage innovation, decision making, problem solving, and initiative throughout the utility rather than isolating it at the top of the organization chart. In short, leadership is an essential element for progressive approaches to managing utilities.

Managers realize that there are numerous forces and values—personal and organizational—that affect their willingness to lead. Everyone, everything, and every situation reflects changing environments, relationships, and expectations. Leaders adjust to those changes and use them creatively to mold a cohesive utility where all resources are best utilized to achieve the shared vision and mission. There are three areas of major importance:

- **The level of confidence a manager has in subordinates.** Leaders must have a high degree of confidence in the capabilities of subordinates in the organization to accept the responsibility and accountability that accompanies the authority to accomplish objectives. This requires a commitment to develop subordinate skills through training, assignment of projects, mentoring, and coaching. It also requires that the manager understands that occasional substandard performance, and possibly failure, may occur as part of the development process. Managers should anticipate appropriate corrective actions.

- **The manager's level of confidence.** Whether it is a new situation, a new process, or new people, a manager knows his or her skills and limitations and takes maximum advantage of each to increase his or her own capabilities. Self-confidence shown by a manager does not present itself as arrogance or bluster but as a presence that is easily recognized and acknowledged by peers, subordinates, and the greater audience.

- **The manager's sense of security in uncertain situations.** Each manager is much more comfortable when in control of a particular situation. The manager has sufficient confidence in him- or herself and subordinates and sufficient knowledge of circumstances and potential events to, at the very least, project a high comfort level in response to uncertainty. Smart managers are good listeners and synthesize information coming from multiple sources. Also, good managers are ready and willing to delay decisions in order to assess information and incorporate it with utility strategies, if that is the most appropriate tactic in an uncertain situation. Finally, good managers establish a reputation for honesty and integrity so that their voice and actions receive attention and respect in the uncertain situation.

Leadership must support the organization's core strategy. The core strategy for most water utilities is operational excellence. Utilities are, by their nature, highly regulated organizations that add value by providing their customers with consistent, dependable, reliable, and cost-effective services. External forces (regulations, customers, resource availability, community mandates, etc.) create the circumstances in which a utility establishes its core strategy, so the organization's culture must be in control of success factors and must focus on obtaining the organization's goals. As with culture, leadership balance is important.

REGIONALIZATION

As resources become scarce and urban expansion occurs, utilities are exploring opportunities to enter into regional partnerships. In some instances, this involves establishing a new agency that involves interested parties to govern, share costs, and benefit from a region-wide project. Regionalization requires commitment from utility partners and related agencies to address resource acquisition, construction, cost, and customer issues on a fair and equitable basis. Several successful examples of regional systems are found in widespread areas. Those that are successful have several common characteristics:

- **They have equitable decision-making processes.** This usually means that all participants have an equal vote in the decisions that affect all parties.

- **They have equitable financial obligations.** During startup processes, this may be an assessment based on a mutually agreed on formula, such as the number of customer accounts or the projected equity in capital projects. As projects come on line, the financial obligations may be based on capital equity or cost of water delivered to the entity.

- **They provide for resolution of disputes.** This recognizes that circumstances and the needs of individual entities within the regional system may change and that there must be a reasonable way to resolve differences without escalating costs or causing the regional group to disintegrate.

- **They provide for the addition of new members.** Successful regional systems recognize that their membership may grow and that new members will want a voice in decisions, as well as access to products and services. They identify an opportunity cost for those additional members and update that cost based on established factors, such as inflation, cost of raw water, construction or expansion requirements, etc.

Leaders of dynamic utility organizations must anticipate change and manage that change with flexibility in policies, procedures, and programs. A key point to remember is that other utility managers have faced similar challenges and opportunities. Excellent leaders have access to a wide network of water utility colleagues who share experience and, often, resources.

In the current utility business environment, managers must be closely attuned to external and internal issues. Governance, ethics, organization behavior, leadership, and vision are equally important to the general public and to utility employees. As resource requirements multiply, managers must exercise collegial behavior in addressing regional needs. All of these elements are found at a time when people live and work longer, values and cultures are transferred more easily, and public scrutiny is more prevalent.

This page intentionally blank.

Chapter **3**

Effective Public Communications

OVERVIEW

Communication is a fragile process, dependent on perceptions, experiences, choices of words, listening skills, and a host of other, often intangible, descriptors. Yet it is the primary strategic tool utility managers must employ in planning, organizing, directing, and controlling the work associated with the organization's mission. And, it is the primary tool used in gaining public support, dealing with stakeholder issues, addressing legislative and regulatory matters, and responding to media and citizen inquiries. Each manager communicates about 70 percent of his or her workday—face-to-face, in writing, on the telephone, listening to others, and reading—in addition to developing and reinforcing his or her own communication skills. Communications is an ongoing process. This chapter addresses the following topics:

- who are the utility's *publics*?

- routine communications

- internal communications

- external communications

- public involvement

- community relations

- effective media relations

WHO ARE THE UTILITY'S PUBLICS?

Utility managers tend to think of the public as a broad spectrum of customers, community groups, regulatory agencies, and policy makers. Managers often forget that one of their most critical and potentially dynamic *publics* are the men and women who work for their utility—the customer service representatives, maintenance crews, meter readers, and others who interact on a daily basis with the external public.

It is imperative that strategic communications planning involves both internal and external publics and that communications are an integral part of all managerial processes, decisions, and actions. Managers must address the information needs of both employees and external audiences. The questions are the same: Who is doing what? Who is responsible? What is planned? What is under way? What is the impact on my job, my neighborhood, or my pocketbook? What does the utility intend to become? Where are projects planned? Where will my job lead me? When will changes occur? When will they affect me? When will promised projects begin? Why are initiatives undertaken? Why do we need additional funds? Why are we implementing new programs? How will new programs be implemented? How will that affect my working conditions? How will that affect my lifestyle or my neighborhood?

The questions and the need to know are legitimate. The need to credibly communicate the utility's message is too often left to chance, but inadequate or inaccurate communications can damage the utility's reputation. A credibility gap appears when there is a disparity between the facts of a situation and what is said or done. Once credibility is lost, it is difficult for further messages to be relied on or accepted as fact. The fragility of the communications process accompanies every manager on a continuing basis—thus, the need to focus on improving skills.

ROUTINE COMMUNICATIONS

Every day, your utility and all its managers and employees engage in communications. Employees answer customer questions; prepare letters, memoranda, and press releases; and participate in meetings to make decisions or learn about new products. The objective is to have common understanding of the topic and to keep the organization's credibility intact when communications are complete. Common threads are courtesy to and respect for all parties throughout the process.

Communications theorists generally agree on several points:

- In general, communications are more spontaneous than planned. Planned communications have better chances of success.

- Most communications are filtered through the perceptions that the audience holds about the speaker or the institution he or she represents.

- Most communicators are inadequate listeners.

A good rule in routine communication, whether it is internal or external, written or oral, is to use the PARTS (purpose, audience, results, technique, and style) process as a planning tool. PARTS is a tool that will improve communications, whether the manager is making a presentation to a board or city council, speaking at an employee meeting, or conducting an individual performance review.

Purpose. All types of communications have a purpose. Whether that purpose is to obtain support for a program or to improve the performance of an employee, there is a reason for the communication. In planning for an information exchange, the manager must clearly understand both sides—why he or she is communicating and why the other group or party is listening and responding.

Audience. It is not enough to know that the audience is from the local chamber of commerce, neighborhood association, or is a problem employee. Learning about an audience can reveal business concerns, lifestyle concerns, and work patterns. It can also lead to solid inferences that better focus communications. Managers must be students of current events. What happens in another city, state, or country is part of the news your audience regularly receives, and it has an impact on their information needs from your utility.

Results. Managers should not engage in communications without having an idea of the results they are seeking. Whether it is support for a rate increase, community involvement in a new project location, or improvement in work performance or behaviors, managers want and need something. So does the audience. Good communications planning requires that the manager define his or her desired results and at least anticipate what his or her audience expects.

Technique. Each communication process requires a different technique. It may be written or oral. It may involve graphics or handouts. It may be formal or informal. The technique selected must match the communications engagement, and the audience must perceive the manager as being fully comfortable using that technique.

Style. Style is where the manager exercises his or her personal power to convey the message to the audience. Style is a critical part of the communications process. Ineffective speakers stand behind a lectern and read a speech with few, if any, efforts to make eye contact with the audience. Effective speakers seldom refer to prepared notes and engage the audience in their message. Each example represents a speaker's style and reflects, to a great degree, his or her comfort level with the topic and reason for speaking.

INTERNAL COMMUNICATIONS

A utility's employees are potentially its best ambassadors, if they are considered high-value assets and if they have the right information at the right time. Utility managers are the primary sources of information, but often managers do not take full advantage of the opportunity to provide information and develop the employees as ambassadors. Internal communications must stress the vision and mission of the utility to all managers and employees and must keep those messages highly visible so they form the basis of all other communications. Internal communications must include both written and face-to-face messages. Above all, internal communications must be a continuous process that is open and honest.

The foundation of an effective internal communications effort is managerial commitment to its success. Many utilities have structured communications efforts that include but are not limited to

- regular briefings of all managers regarding the utility's programs and issues
- regular employee newsletters
- functional staff meetings
- all-staff retreats and work sessions

The Seattle Public Utilities Communications Office developed a comprehensive communications program based on the belief that establishing clear, effective communication with the public, the news media, and the utility's employees is a vital strategic priority. Its internal communications efforts are based on the following principles:

- ensuring that employees understand and support the utility's objectives and that they are motivated and aligned with its goals

- making communications a tool used to achieve a competitive edge

- tailoring messages to empower employees to make better decisions

- building on the interest and direct relationship that all employees have in the utility's success

- managing and disseminating information in ways that enhance the intellectual capital of the utility and its employees

- expecting employees to positively represent the organization by equipping them with knowledge so they respond to inquiries with factual information

- providing consistent, uniform information that mitigates rumors and misinformation

- getting the right information to the right people at the right time

EXTERNAL COMMUNICATIONS

External communications include regular contact with elected officials, community leaders, key customers, neighborhood groups, the media, and civic associations. It also includes ongoing communications with regulatory, governance, and financial agencies. Written communications, such as press releases, quarterly or annual reports, newsletters, or other information that appears on bills, are the most frequently used. Television and radio news organizations often report on community-wide topics, such as rate increases and major construction projects, or service disruptions, such as major main breaks. It is important that communications efforts be considered strategic efforts and part of an overall plan to advance the utility's mission within its community.

Many utilities take advantage of relatively inexpensive ways to get their messages to their various external audiences. Videotape programs on subjects such as water quality, water resources, utility expansion, environmental quality, and conservation, are often produced by utility staff, local cable television systems, or public relations consultants. Such programs can be checked out to employees, offering them a way to take the utility's story to various audiences, like clubs, school groups, and neighborhood associations, and become ambassadors for the organization. There are often public service requirements that radio and television stations must meet or that are part of a cable system's franchise documents.

In addition, expanding technology and utility Web sites offer customers an opportunity to monitor governance meetings, obtain news about utility activities and programs, learn about service policies and rebates, ask questions, review consumer confidence reports, and view some of the utility's work activities or programs as outlined above. Web sites can be interactive and allow for bill payment, service requests, information inquiries, etc. Utilities are cautioned that Web sites require an ongoing commitment to update materials—they cannot be left unattended—and ongoing attention to security requirements.

A continuing question is whether communications efforts should be outsourced or staffed internally. Because communications are part of the utility's strategic direction, it is logical that an individual on the utility's staff is designated as the communications officer. This person must be known to external agencies as the source for information and must have sufficient time and credibility to represent the utility when called on. In many organizations, customer service and external public relations are combined, while in others, external public relations are the purview of the general manager. Given the critical nature of a consistent and professional communications effort, many utility managers recognize the value of having a dedicated staff member responsible for setting and implementing strategic communications. Some utilities choose to contract with external resources for public relations services on an as-needed basis or for specific projects or critical issues.

All utilities must prepare and distribute a Consumer Confidence Report to their customers, and many utilities prepare an annual report of their activities and challenges. These types of publications can provide consumers with vital information that builds the utility's credibility, as well as give the utility managers a way to promote the utility. Such reports do not have to be expensive, but they do require a professional appearance, clear language, and meaningful graphics. Above all, the content of the report must be truthful.

The underlying principles that govern external communications efforts include but are not limited to

- establishing clear and effective high-quality communication strategies, programs, and materials that are consistent with the utility's mission, goals, and values

- involving the communications team in all programs, projects, and plans at the beginning, not when the programs, projects, and plans are in trouble. communication advice and strategies cannot be an afterthought

- implementing a timely and efficient review process for all materials intended for broad public and employee distribution to ensure that the materials are consistent in style and that they are courteous and positive in their wording. this includes all editorial items, such as newsletters, news releases, flyers, brochures, speeches, reports, scripts, and displays that represent the utility. it also includes promotional and advertising materials produced by external agencies

- coordinating all requests for media interviews, information, presentations, speeches, etc., through the communications center so organizational objectives are incorporated and quality and consistency are maintained

PUBLIC INVOLVEMENT

Public involvement is a permanent part of water utility management. Managers must seek public involvement and actively listen to the needs, expectations, and concerns of all audiences. Public involvement in utility decisions is essential to building and maintaining credibility and public trust. The utility manager should recognize public involvement as a crucial asset to the organization's ultimate success. It is the utility manager's responsibility to manage the public involvement process.

Some utility managers question whether—or how much—public involvement is necessary. Other utility managers question whether the value gained from public involvement is significant in light of the required time, dollars, resource materials, etc. Utility managers should recognize that their customers want to be involved in and kept abreast of utility plans and projects and that the public desires to make appropriate contributions. Indeed, if there is a discrepancy, it is that the public's desire to participate in utility matters is significantly greater than managerial awareness of the need for involvement. Many utilities successfully embrace a public involvement strategy while others test the water.

Engaging in broad public involvement requires the utility's leadership to answer several questions:

- Do utility managers value external input?

- What experiences have utility managers had when public or stakeholder groups have been involved previously?

- Do utility managers trust in their ability to manage the public process?

- Can the results of the public process lead to an internal or external crisis of confidence?

Too often, the source of information about what the public wants, needs, and is interested in comes from traditional interest groups—business leaders, developers, and government agencies—and does not include other stakeholders, including neighborhood associations, environmental groups, and neighboring communities. These groups form coalitions to make their opinions known. Utility managers are challenged to understand the needs of all stakeholders, respect them, and communicate in a collaborative way to build the broadest possible understanding among all groups.

There are several ways to determine public interests and public concerns. The easiest way is to ask, such as at a meeting where internal and external communications strategies can be presented. Utility employees are arguably closer to a more diverse portion of community than public information experts. Employees are a rich source of information, but managers have to make time to talk with them, engage their assistance, and receive feedback. Other established methods include

- providing avenues for two-way communication, such as citizen advisory groups, semiannual meetings with key customers, regular dialogue with neighborhood groups, listening sessions with stakeholders and employees

- commissioning professional market research and customer surveys, and analyzing the information to determine perceived strengths and weaknesses the utility can correct or opportunities the utility can take advantage of

- establishing relationships with local news reporters and talk show hosts and using those contacts to disseminate information to provide accurate responses when facing adverse conditions

- consulting and sharing information with other utilities in the area through regular meetings and interlocal agreements for assistance

- accessing industry organizations, such as awwa, and networking with members

COMMUNITY RELATIONS

Community relations are an extension of public outreach and a predecessor to public involvement. Utilities have numerous stakeholders or communities they must address and satisfy. Effective community relations begin with a comprehensive understanding of the stakeholder groups and their needs and are accompanied by a plan to manage the information required by each. Generally, stakeholder communications involve meetings, whether in neighborhoods or with special groups. Public involvement and community outreach programs require the commitment of resources and adequate staff to monitor and manage the effort.

Figure 3-1   Successful communication with the public is essential for a water utility's success.

It is important that any type of community outreach effort have a clearly defined purpose and process. Its purpose must be stated and frequently revisited. Managing the process requires open and balanced communications among all parties, including rules of courtesy, shared responsibility, agreement on mutual commitments, patience, and a willingness to stay with the established purpose and goals. A key element in successful outreach is a commitment to factual information and reasonable actions, not speculation and impossible promises. Establishing a follow-up plan to continue dialogue and close any open issues emphasizes the utility's credibility by establishing the outreach effort as an ongoing process.

It is also important that presentations to community groups be made in understandable language and, where possible, accompanied by graphics that enable participants who do not know much about the water system to relate to the information. Technical presentations are suitable for knowledgeable groups, but technical terminology is confusing to nontechnical audiences. Photographs, easy-to-understand charts and graphs, and common language are powerful tools in delivering the utility's message. Those who make presentations must have the skills to do so, including making eye contact with the audience, being comfortable before an audience, having familiarity with the subject matter, and having an ability to convey mastery or understanding of the topic or issues without being patronizing or condescending.

EFFECTIVE MEDIA RELATIONS

Effective media relations are essential for telling the utility's story. Effective media relations cannot be initiated on the day of a crisis. Like all other parts of the communications process, relationships must be established and cultivated beforehand.

A utility builds credibility with the media by using the same principles and programs it relies on to build credibility with customers and employees. Utility managers must

- engage in proactive and open communications

- know who needs to hear what and when

- seek external expertise when needed (advisory boards are useful and can help to build understanding and support)

- reveal what information is needed before being asked

- adopt a cooperative posture when dealing with media representatives

All utility leadership must know and espouse the organization's key messages in all public commentary. Communications professionals assist in developing appropriate tools and formats and in identifying the target media outlets, but it is up to utility leadership to remain focused on key messages in their commentaries. It is also advisable that the utility have one spokesperson who is well trained, has access to decision makers, and focuses on establishing and maintaining contact with media representatives. The media—and all utility employees—must be aware that the spokesperson has a straight line to policy makers and senior management and must be immediately involved if there is a request for information. No utility wants or needs to have multiple spokespersons or the resulting inaccuracies or distortions that often emerge with more than one spokesperson.

Inaccuracies and distortions result in loss of credibility, and once a utility has lost credibility, it takes years to rebuild. Credibility is evaluated through honest self-evaluation and in the public's response to media reports. When faced with negative publicity, it is easy to fall back to the observation of inaccurate reporting. However, those inaccuracies may result from lost credibility. If the answer to any of the following questions is yes, the utility may already have lost some credibility:

1. Has the utility received negative publicity and have utility managers blamed the bad press on inaccurate or biased reporting?

2. Has the utility ever corrected a problem but had little or no followup with those who were impacted by that problem?

3. Have utility spokespeople ever resorted to a technical justification for failure rather than calling it an error?

4. Have utility managers ever been unavailable to respond to the hard questions that customers or the media ask?

There are some basic rules that apply to media relations, and these should be remembered when developing and sustaining effective media relations. One of the first rules is to be friendly with media representatives. Part of that friendliness is to have background information about the utility readily available for media representatives; press releases should address basic questions—who, what, when, where, why, and how. The background material might include stock photographs of utility activities or personnel, such as utility employees repairing the distribution system, a laboratory technician conducting testing, or customer service representatives taking customer calls. Print and electronic media reporters have deadlines and the utility gains credibility when it knows and respects those constraints. Press conferences that are scheduled 30 minutes before those deadlines do not engender goodwill. Briefing sessions on major issues, events, or projects should be timely, and utility managers should be prepared for media inquiries in the wake of drinking water problems in neighboring localities.

Many utility managers have taken advantage of media training to hone their knowledge and skills; but to improve interactions with the media, utility managers need to follow certain basics:

- Know how the media works—who they are, what they do, what they want, how they operate, and what their situation is.

- Be interested in the topic at hand and convey genuine concern.

- Take the opportunity to deal with the media, but be prepared and honest. Managers cannot accept the premise put forth by the reporter as the whole truth and should take time to formulate an answer. Managers must tell the truth, even if it brings negative consequences. If the answer to a question is unknown, managers should state that they do not know but will try to find out and follow through.

- Control the situation. Managers can simplify the complicated by explaining their message so people understand it. Above all, managers must remain calm.

- Prepare for the interview by reviewing relevant files, making an outline of important points, and learning the answers to who, what, when, where, and why. Managers should also prepare background materials, ensuring that the information presented is timely and accurate.

- Take time to review the subject with the reporter before the camera or tape recorder is turned on and get an idea of what questions are to be asked.

- Always assume that everything said is being recorded—whether in person or in a telephone interview.

- Be aware that there is no such thing as *off-the-record*, and a statement of "no comment" often raises a credibility issue. If a manager cannot or is unwilling to answer a question, he or she should state this and give a reason why.

- Prepare for a crisis and avoid creating an unnecessary one. If managers do not want something reported, it should not be addressed.

Communications—whether internal or external, crisis or routine, good news or bad—is the glue that binds the utility staff and its publics. An informed staff is a motivated staff; an information-centered manager is a respected manager whose credibility and honesty is highly valued.

This page intentionally blank.

Chapter **4**

Utility Financial Management

OVERVIEW

Since January 1965, the American Water Works Association's board of directors has had a policy statement on utility financing and rates. From time to time, this policy statement has been revised and reaffirmed by the board. The policy, as revised in 2005, states:

> The American Water Works Association (AWWA) believes that the public can best be provided water service by self-sustained enterprises adequately financed with rates and charges based on sound accounting, engineering, financial, and economic principles.

> To this end, AWWA recognizes the following principles that water utilities should establish. Implementation of these principles can be balanced against other policy objectives; however, no policies should be adopted that compromise the long-term financial integrity of water utilities or their ability to provide service to customers. Basic financing and rate principles include:

> 1. Water utilities' revenues from water service charges, user rates, and capital charges (e.g., impact fees and system development charges) should be sufficient to enable utilities to provide for:

>   • annual operation and maintenance expenses;

>   • capital costs (e.g., debt service and other capital outlays); and

>   • adequate working capital and required reserves.

> 2. Water utilities should account for and maintain their funds in separate accounts from other governmental or owning entity operations. Water utility funds should not be diverted to uses unrelated to water utility services. Reasonable taxes, payments in lieu of taxes, and/or payments for services

rendered to the water utility by a local government or other divisions of the owning entity may be included in the water utility's revenue requirements after taking into account the contribution for fire protection and other services furnished by the utility to the local government or to other divisions of the owning entity.

3. Water utilities should adopt a uniform system of accounts based on generally accepted accounting principles. Utility practices should generally follow the accounting procedures outlined in the water utility accounting textbook published by AWWA. Modifications may be made to satisfy the financial and management control reporting needs of the utility and to meet the requirements of legislative, judicial, or regulatory bodies.

4. Water rate schedules should be designed to distribute the cost of water service equitably among each type and class of service. Non-cost of service rate-setting practices may be appropriate in some situations, subject to legal review and approval, provided they reflect market conditions, the benefits received by the users of the service, and an appropriate balance of the goals and objectives essential to the public good. Any non-cost of service rate-setting practice implemented by a utility should be fully disclosed to its customers, regulators, and the financial community. Such disclosure should identify each non-cost of service rate-setting practice, its expected benefit, and its impact on the utility's customers.

5. Water utilities should maintain asset records that detail sufficient information to provide for the monitoring and management of the physical condition of infrastructure. These asset records should also support planned and preventive maintenance programs and budgets adequate to maintain the utility's assets at a level of service consistent with good utility practice. Utilities should annually provide comparative information to customers, the financial community, and the general public about the utility's sustained capability to provide water service and generate revenue levels necessary to protect the financial investment of others in the utility. Such information could include historical expenditures for renewal and replacement during each of the past several years, as well as the revenues that would be generated under planned and adopted rates to support renewal and replacement during each of the next several years.

Key players in the utility's financial picture are the governing organizations it reports to or liaises with, regulatory agencies, its financial partners, and its customers. Ongoing communication is needed with elected officials, organizations, and agencies whose distinct roles and responsibilities lie in the utility's service areas so that coordination and collaboration enhances available funds for projects with minimal disruption of service and customer dissatisfaction.

FUNDING

Utility managers have numerous sources of funding, including revenue bonds, grants, state revolving funds, special appropriations, public–private partnerships, private investment, and funding for targeted service areas, such as rural, unserved, or underserved populations. Access to such funding and how the utility best applies those funds to meet its service mandates are important elements of a manager's responsibility.

Rates and Charges

Comprehensive information about developing cost-of-service rates and rate design are discussed in detail in AWWA Manual M1, *Principles of Water Rates, Fees, and Charges*. The overall concept of rates based on cost of service is that the rates for a particular class of customers should reflect the cost of serving the customers in that class. The cost of service for a particular class of customers is based on allocations that recognize, among other factors, the number of customers, the class average and peak demands in proportion to the total number of customers, and system average and peak demands of the utility. As a result of the matching of costs and revenues by customer class, cost-based rates are generally considered both fair and equitable. Variations from cost-of-service–based rates to meet specific policy or practical considerations of the utility, ranging from affordability concerns to industrial incentives, are accommodated within the rate-setting process. It should be recognized, however, that the total revenue requirement must be recovered from the utility's rates and charges in an enterprise fund or self-sustaining utility environment.

Rate Design

While there are broad principles governing rate design, alternative rate designs applicable to water utilities have emerged over time. Some have been driven by the historical and often regionally based evolution of different rate designs, while technology (metering), issues regarding equity, concerns about water supply and quality, and other management objectives have influenced others.

Rate schedules for water service typically include two types of charges. The first is a *fixed fee* per billing period that usually does not vary with water consumption. The second is a consumption charge that usually varies proportionally with the quantity of water consumed during the billing period.

The fixed fee is usually designed in one of three forms:

- A *customer charge* is a fixed amount that is the same for all customers.

- A *meter charge* or *service charge* is a fee that increases with the size of a customer's meter.

- A *minimum charge* may be either a customer charge or a meter charge that was increased to include the cost for a minimum amount of water consumption. This consumption allowance sometimes varies with the size of a customer's meter.

Consumption charges generally fall into one of four categories.

- The *uniform consumption charge* prices all water at the same unit price.

- The *decreasing block consumption charge* has had widespread use. This design divides a customer's consumption into blocks (i.e., ranges of consumption) and charges a lower unit price as a customer's consumption increases and moves from one block to another block. In this case, a customer's bill is usually calculated based on the consumption in each block multiplied by the unit price (rate) for each block. If a declining block rate structure has been established appropriately, it reflects the cost of service of different customers consuming in various usage blocks.

- The *increasing block consumption charge* has increased in popularity because of its ability to support conservation objectives. This rate structure focuses on conservation by dividing a customer's consumption into various usage blocks and charges a higher unit price as a customer moves from one block to the next. Like the decreasing block consumption charge, a customer's bill is generally calculated based on the consumption in each block multiplied by the unit price (rate) for each block. Excess-use rates form a variation on the increasing block rate structure as increasing rates tailored to an individual customer's usage characteristics.

- *Seasonal rates* are also conservation focused and charge different unit prices during high-peak and low-peak demand periods.

Related Charges

There are other types of charges for water service, beyond the basic water rate structures discussed above, that a utility may assess to individual customers based on their specific requirements for service. These include *system development charges* (SDCs) or impact fees, *connection charges, line extension charges,* and other *miscellaneous fees*.

Generally SDCs are a fee assessed to new customers or new demands placed on a water system. The fee is based on the cost of providing system capacity in source of supply, treatment, pumping, and transmission capacity for the new customer and is intended to insulate existing customers, to the extent possible, from the cost associated with system growth. Connection charges are fees that are specific to a new connection to a water system and generally recover the cost of the service connection and meter. This fee does not generally include the cost of the system capacity in supply, treatment, pumping, and transmission. Line extension charges are also generally handled on a case-by-case basis and are intended to recover the cost of extending a service to a new area. Miscellaneous fees and charges for services, such as special meter readings, turn on/turn off, returned-check charges, cross-connection inspection, and other special requirements, are recovered directly from the customer needing the service.

It is essential that the utility regularly review rates and charges, as well as water use and sales information. Revenue may become inadequate because rates combined with sales do not generate revenue that keeps pace with rising costs. Regular reevaluation of revenues and costs allows the utility to take early corrective action in cost controls and to lay the groundwork for additional revenue alternatives. This is typically accomplished through the long-range financial planning process.

Capital Financing

Water rates and charges should be established at a sufficient level to fund not only the ongoing operations and maintenance requirements of the water utility, but also capital costs (such as debt service requirements and routine capital outlays), required reserves, and adequate working capital. Financing sources for water utility capital expenditures include

- water revenues, such as service and user charges
- capital charges, such as impact fees and system development charges
- debt issuance, such as revenue bonds
- commercial paper obligations
- other sources or charges

Comprehensive information regarding capital financing options can be found in AWWA Manual M29, *Water Utility Capital Financing*.

The governing board determines to what extent any capital financing option is used by the water utility, based on the recommendations of management and, in some cases, a financial advisor to the utility.

Capital planning includes assessment of growth history and projections, land-use patterns, assessment of current system capacity, and the planning horizon. Other areas of consideration include new facilities and the expansion of current facilities, replacement and rehabilitation/reconstruction project requirements, budget costs, project priorities, and project timelines. Financial considerations include current and projected revenues and expenses, reserve requirements, and projected surplus or deficits.

FUNDING ALTERNATIVES

Each funding agency has specific reporting requirements for funds expended, their use, and project status. It is essential that the utility implement and maintain financial control mechanisms for and with all functional units associated with any project and that all documentation is approved and maintained according to established procedures. Regular internal financial reporting, accompanied with appropriate documentation, should be completed according to a time schedule agreed on at the beginning of any capital project. Not only does this process expedite payment for capital project costs, but it is also important in meeting requirements of an annual audit.

ACCOUNTING

Water utilities should adopt and maintain a uniform system of accounts based on GAAP (*Water Utility Accounting, 3rd ed.*, is a good reference on water utility accounting). Under Governmental Accounting Standards Board (GASB) requirements, the modified accrual basis of accounting is specified for enterprise fund accounting (www.gasb.org/). The GASB Statement 34, commonly called GASB 34 (www.gasb.org/repmodel/index.html), requirement modifies how financial information is presented in the audited financial statements of the utility. It also stipulates a *Management's Discussion and Analysis* section in the annual audit.

GASB 34 provides an alternative accounting and reporting treatment for repair and maintenance expenditures that may be more reflective of the maintenance programs pursued by some utility managers. This accounting treatment is referred to as the "modified approach" and is described in paragraphs 23 and 24 in the GASB 34 statement. If the prescribed accounting treatment under the modified approach is more reflective of the utility's repair and maintenance program, discussion with the chief accounting and finance officer to implement this alternative approach should be pursued. Extensive guidelines were prepared to assist the utility (and other government units) in addressing GASB 34 requirements. An independent auditor should be contacted for a more detailed discussion of the various GASB 34 requirements.

Financial- and accounting-related reports should be generated by utilities throughout the budget year. Common reports include comparisons of actual versus budgeted expenditures for operations, maintenance, and capital expenditures. Utilities should also generate monthly or quarterly unaudited financial statements throughout the budget year. Additional reports may be generated for other utility stakeholders, such as regulatory agencies and lenders. The type and detail of these reports vary by end user. For example, reports generated for the governing body are typically less detailed in nature than those generated for utility staff in charge of

Figure 4-1 The public can be best provided water service by self-sustained enterprises adequately financed with rates based on sound accounting, engineering, financial, and economic principles

managing departmental budgets. The main purpose of these reports is to manage the ongoing financial obligations of the utility, identify budgetary concerns (operating and capital) throughout the year so that necessary adjustments are made, provide financial information to internal and external stakeholders, and facilitate corrective action as necessary.

MANAGEMENT CONTROLS

Each funding agency has specific reporting requirements for funds expended, their use, and project status. It is essential that the utility implement and maintain financial control mechanisms for and with all functional units associated with any project and that all documentation is approved and maintained according to established procedures. Regular internal financial reporting, accompanied with appropriate documentation, should be completed according to a schedule agreed on at the beginning of any capital project. Not only does this process expedite payment for capital project costs, but it is also important in meeting the requirements of an annual audit.

BUDGETING

Each utility should develop an annual operations and maintenance budget and a longer-range capital budget (if capital improvement projects are planned) for adoption by its governing body. Budgets must be adopted prior to the beginning of a fiscal year, thus the budgeting process begins several months in advance of the adoption date. A budget is an officially approved expenditure document. The operating budget is usually set for one year, in part because one governing board cannot commit a future board to an expenditure program through the budget process.

Although biennial budgets are becoming more commonplace, a governing body may approve a five-year capital improvement program each year, but expenditure authorizations are only made one year at a time. Prudent financial management suggests that a utility develop a long-range financial plan covering a comparison of revenues and expenditures for several future years as a guide for financial matters. This financial plan could be systematically implemented each year through the budget process.

Utilities adopt different processes by which they budget. Some involve a citizens' advisory committee that assists in preparing the budget and may, in some cases, actually present the budget to the governing body. In most utilities, however, the senior management staff of the utility is charged with the budget process.

- For operations and maintenance budgets, managers are usually provided a copy of the prior year's budget categories and approved funding allocations. This information is usually accompanied by a year-to-date spending level so the manager can compare his or her original projections with actual expenditures. Budget packages may include forms to document new programs and their costs (personnel, materials, tools, equipment, etc.). Increases to budgets usually require written explanation and supporting cost data.

- Capital budgets may be more restricted, especially if bond funds are used and the bond offering has specified the projects to be completed. Managers have a responsibility to evaluate capital requirements and to list and prioritize them based on system-capacity requirements. System inspections by regulatory agencies identify areas where significant capital projects are needed—additional wells, additional ground or elevated storage, treatment process upgrades—and this information becomes a part of the capital program budgeting process. When funds allocated to a particular project are not fully used, the utility evaluates its capital program listing and selects projects to be funded. Many times, the capital budget is an outgrowth of a utility's master plan.

A comprehensive draft budget reflecting all anticipated or desired expenditures—coupled with projected revenues from water sales and other revenue sources—should be prepared and circulated among the utility's senior managers, followed by budget negotiations among leaders and the utility's financial staff. Information generated in the budget negotiations, in most cases, is translated into necessary increases in water rates or fees and a final budget document and rate resolution are developed for approval by the utility's governance body.

In many utilities, the governing body's finance committee participates in the budget process, providing direction to the managers as to what the community or the service area needs in the way of services or infrastructure improvements. The inclusion of the finance committee provides utility managers with an opportunity to explain operations, identify areas of improvement that require funding, and build support for the budget and rate documents. It also provides an opportunity for a frank dialogue about financial aspects of the utility and an environment where concerns about costs and expenditures can be voiced and discussed.

Some utilities hold public workshops and public hearings about their budgets, inviting questions and engaging in dialogue with concerned citizens about costs and charges. In general, public hearings serve their purpose best when the utility has built a strong reputation of water quality and reliable service and when its public communications are ongoing and viewed as highly credible.

AWWA offers a variety of manuals, financial seminars, and other resources to assist the utility manager in meeting fiscal responsibilities. Additionally, many public and municipal utilities take advantage of programs offered by the Government Finance Officers Association.

Chapter **5**

Customer Service

OVERVIEW

Delivering quality water to customers is a utility's core business, but delivering excellent customer service is what makes a utility successful. Excellent customer service means the utility exceeds customer expectations. The frontline staff are a manager's eyes and ears and provide a window to the customer. Utility personnel reflect management's customer service philosophy, vision, values, and goals. Whether an employee works in the customer service department, in the lab, at a treatment plant, or in field maintenance and construction, they have customers and are customers. The first contact is where customer opinions are formed, regardless of whether the customer is external or internal.

Water utility managers are growing more aware of and placing greater emphasis on quality customer service. Customer service has always been a major component of a utility manager's business model. Public expectations of transparency in governance and for immediate and exceptional customer service are the norm for most

Figure 5-1 Delivering quality water to customers is a utility's core business, but delivering excellent customer service is what makes a utility successful

utility operations. It is imperative that utility employees—from the highest level of management to the newest work team member—are aware that customer service is a part of their job.

In a practical sense, this means that utility leaders make a commitment to exceptional customer service and set the example by establishing and exhibiting the utility's belief that the customer is its highest priority. It also means that all employees of a utility are responsible for establishing and maintaining positive working relationships with the public, coworkers, and any specialized groups or individuals the employees come in contact with. All employees must know the utility's service goals and how they fit into the overall vision and values of the utility. These goals are not a one-time message—they must be repeated frequently in meetings, in written materials, and through any other appropriate methods. Employee orientation must emphasize the customer-first philosophy, and additional courses to develop customer-relations skills are frequently a component of on-the-job training programs.

Utility managers and employees must recognize the various types of customers served and must accommodate the different needs of those customers. Utilities profit from heightened employee responsiveness tailored to residential, commercial, and industrial customers, as well as to developers and contractors. Each type of customer requires different types of service and each has different expectations of the utility. Utilities must be adept at recognizing the differences and responding accordingly.

For example, younger customers are often more technologically savvy and expect the same type of customer service they receive from other utilities and service providers. Online billing, payments, and account status reviews are a normal way of doing business, and they expect the utility to provide that level of service. Conversely, older customers may still consider it important to present their payment to the utility in person, and they expect a customer-friendly service center that is easily accessible by available forms of transportation. Customer-service operations must be aware of critical-needs customers—for example, hospitals, life-support systems. This information is essential to meeting needs during service interruptions, whether scheduled or unscheduled.

A water utility's customer service operations must consider the regulations that govern its functions. Regulations vary, governed by national, provincial or state, and municipal requirements. Of particular importance are the regulations dealing with billing collection, delinquency, and the use and dissemination of information.

All utilities recognize the importance of external customers—people who have accounts with the organization and to whom the utility provides water and wastewater services. Historically, the customer service emphasis has focused on external customers. However, most utilities also recognize the need to improve internal customer service—the information shared with, and responsiveness to, the needs of other employees and work groups within the utility so that they are better able to meet their objectives and the needs of their customers. A manager's responsibility for the customer service function(s) and climate within the organization is discussed in this section.

One simple way of emphasizing the importance of each employee is to provide all new employees with a business card. Business cards are produced easily and inexpensively using a word-processing program. It should carry the utility's logo, address, telephone number, and the employee's name. The advantage of using this approach is twofold—it gives the employee the sense that he or she is important enough to the utility's work team to merit a business card, and it allows the employee to respond to a customer's inquiry by giving them an immediate way to contact the utility and be directed to the appropriate service unit. An underlying foundation of this approach is that it complements a customer-focused management philosophy practiced in all utility functions. Employees need to know that managers expect them to use their skills to help customers. This means that managers must stress customer service training and the sharing of information to allow staff to interact confidently with their customers. Managers must encourage employees to learn without fear of being reprimanded for occasionally making mistakes.

FUNDAMENTAL PRINCIPLES

Water services consist of the connection or service lateral between the utility's water main and the customer's service address. Services are usually through a water meter. AWWA's policy on metering is that all water services should be metered and usage measured to ensure proper customer billing for water used. Meters provide a clear demarcation between utility-owned property and customer property. For many utilities, all infrastructure leading to the inlet side of the water meter and the meter itself are the property and responsibility of the utility. Other utilities consider their infrastructure property and responsibilities to end at the curb (corporation) stop, regardless of the location of the meter. All piping connecting to the outlet side of the meter is the property and fiscal responsibility of the customer.

This information—possibly including a drawing—should be part of the information provided to customers when they apply for service with the utility. The utility also provides the customer with any service policies that include the customer's responsibilities and a rendition of a water meter and how to read it. Including a sample bill teaches the customer what to look for in terms of consumption. This addition gives the utility the opportunity to highlight billing for other agencies that may appear on a monthly invoice and to remind the customer that the water bill includes other services in addition to water.

Different utilities may apply special charges and fees approved by their governing body or a regulatory agency to provide a funding source for capital installations, improvements, and equipment necessary to make water service available. Many utilities favor a rate structure that collects sufficient funds to provide for capital improvements, covering the costs of system expansion as well as ongoing maintenance and operations. Regardless of the billing presentation, it is essential that the utility take every opportunity to explain the bill to its customers and to ensure that they understand the bill's components.

As an example, a utility's billing statement might be "water service," which encompasses a readiness-to-serve charge, a basic consumption charge, a conservation rate charge at a higher rate, or special fees. Customers often look at a generic charge and do not understand the content and believe they were overcharged. This misunderstanding often leads customers to call the customer service center with a complaint that could have been avoided with better delineation of the billing components.

Customer classes differentiate among residential, commercial, industrial, and multifamily locations. In most utilities the costs of installation, fixed-user charges, and consumption are based on the type of customer, the size of meter or the number of fixtures within the premises.

Many utilities serve critical industries or facilities, such as hospitals, schools, major employers and businesses, and government locations. A useful strategy for building community support of and credibility with the utility is to identify these key customers and assign a team of representatives as their primary contact for inquiries or assistance. A typical key customer team consists of a manager, a knowledgeable operations representative, and a customer service representative. The team ideally meets with the key customer at least twice each year with two attendant goals:

1. To determine how the utility might help the key customer do business more efficiently and effectively.

2. To share with the key customer the utility's overall plans for improving service and obtain feedback about how it is meeting the customer's overall and specific needs.

While smaller utilities may not be able to dedicate the resources to this type of approach, it is still important that key customers be recognized in some manner by the management team. Alternatives, such as periodic telephone calls, program update newsletters, etc., can give key customers extra attention.

KEY CUSTOMER SERVICE FUNCTIONS

Almost all utilities, regardless of size, perform certain essential customer service functions or work processes, including

- reading customer meters on a regular basis to determine usage

- calculating, creating, and delivering bills to customers

- processing customer payments

- opening, closing, or transferring customer accounts

- responding to customer inquiries

- investigating and resolving high- or low-consumption inquiries and other customer-related problems

- pursuing arrears and collections or liens for customers who fail to pay their bills

- establishing new services

These various functions may be located in different units in different utilities. Most critical to the utility manager is that all units are aware of their customer service responsibilities and that there is continual communication between units to maintain that awareness at a high level. Among the typical customer service work processes are meter reading, field service, billing (including pre-bill audit), remittance processing (including cashiering), customer contact (including the call center and counter services), office collections, field collections, field meter service, meter repair, and new service installation.

Each customer service process consists of a series of tasks or steps that should be the focus and responsibility of utility managers. A utility manager needs to periodically review such work process elements as

- *Load:* How many times is the process performed (e.g., the number of high bill complaints the utility receives each month and from what service areas)? What is the distribution of the load over time?

- *Resources and productivity:* Who works on the process? What is their productivity (e.g., the number of tasks handled per day)? Are staff properly trained, motivated, informed, empowered, and supervised? Are there sufficient resources applied to the process at the right times or are there bottlenecks?

- *Cost:* How much does it cost to execute the process for each transaction, including labor and benefits, transportation, equipment, overhead, etc.?

- *Reliability:* What percentage of the items flowing through the process are completed as opposed to incomplete?

- *Elapsed time:* How long does it take to process a customer request (e.g., an adjustment for a high bill)?

- *Underlying policies or procedures:* What credit information is required of applicants for new accounts? How many estimated reads does the utility allow?

- *Quality:* How well are tasks in the process handled? How satisfied are customers by the process?

CUSTOMER SERVICE POLICIES

Each utility should establish and make public its policies for providing service. These policies and their attendant procedures should clearly explain the utility's responsibility and the steps it takes to implement those policies. In this way, consistency in responding to customers' service requests is established and the utility gains a reputation for fair and equal treatment of all customers. Examples of topics for policies and procedures include but are not limited to

- application for new service
- water and sewer connection protocols and fees
- billing adjustments
- customer responsibilities
- theft of water
- collections
- termination of service for nonpayment
- appeals
- payment options

Other policies and procedures are needed to address specific local conditions. It is important that all employees who work with customers be familiar with the policies and procedures, that they document all customer inquiries (both for easy reference and to identify potential areas for services improvement), and that clear definitions are in place regarding the scope of authority service representatives have to resolve a problem.

CONSUMER CONFIDENCE REPORTS

Each utility is required to annually provide its customers with a report concerning the quality of water delivered to their home or business. The technical content of the Consumer Confidence Report (CCR) identifies how the utility's potable water compares with the maximum contaminant levels established by national and state or provincial regulations. This provides consumers with a tool to judge the quality of the water delivered. CCRs offer the utility the opportunity to provide customers with additional information—growth, new system components to improve service, conservation messages, etc. While CCRs are only required on an annual basis, establishing and maintaining ongoing contact with customers reinforces the message of providing quality water when needed. Billing inserts or customer newsletters often carry important information about the utility's operations, as well as periodic quality information.

Water quality issues may arise, requiring the utility to issue boil-water notices or other cautionary advisories. From a customer service standpoint, it is helpful if the utility assists the customer in meeting any immediate drinking water needs while providing the best possible information as to the length of time customers may expect the boil-water notice to be in effect. Service personnel must know how to handle customer concerns about the safety of the water as well as the procedures the customer must follow to ensure that water from their tap is safe to use.

CUSTOMER INTERACTION

There are many types of customers, and each type has different needs. They expect a variety of methods to do business with the utility. The size of the utility and budget determines the complexity of systems used to serve customers.

Utility managers have the opportunity to dramatically improve both customer service and employee productivity. The following are some tools available to managers:

- remote meter reading systems that allow meter readers to quickly capture readings electronically or automatic meter reading systems that transmit readings over radio or telephone

- customer information systems (cis) that handle all major customer service functions and customer relationship management systems that maintain and analyze all customer transactions

- advanced telephone systems that streamline customer service transactions and, through integration with cis, enable customers to obtain information about their accounts

- portable computing devices that enable field service employees to carry and electronically capture information in the field

- wireless data systems that send and receive information from field locations

- internet and e-commerce solutions that enable customers to see detailed information about the utility, their consumption, and their accounts; pay bills; and conduct other customer service transactions

Some of the following described key elements are common to all utilities.

Telephone Systems

Most customers who interact with the utility do so by telephone. To them, your customer service representative is the utility. Whether a utility's telephone system includes few lines, an automated call distribution center, or interactive voice-response systems and computer-telephony integrated systems, customers want access to a courteous customer service representative in a timely manner. In choosing a telephone system, it is important to determine the needs of both the customer and the utility. Some utilities must answer calls on a 24-hour basis, take voice mail messages, and direct customers to other frequently called areas of the utility. They need to provide information to customers, such as office locations, hours of operation, and emergency numbers. Others have customers who want to access their account information in order to pay bills. Telephone systems provide utilities with solutions for managers, but they must determine customer needs first. Taking a survey of utility customers often yields valuable information to help determine their needs.

Information Systems

Most utilities use automated information systems to help trace information about customer requests, field work orders, meter sales, meter reading data, billing, etc. The flow of information begins with a request being made by a customer (internal or external) to perform a customer service function. Tracking begins when the request is made and follows the work through the appropriate processes. Information generated assists in refining business processes. It is important that cross-functional teams of the utility's best and brightest employees are involved in implementing new systems and that all employees understand that the changes are the foundation of a customer-focused utility.

Information technology is an evolving component of modern utility management. It is important that utility managers establish and adhere to a recurring assessment of information needs so that the resulting systems address their business needs and the chosen hardware and software are appropriate for the utility. Several steps assist in the process:

- reviewing (usually annually) the utility's business direction and needs

- conducting internal and external assessments of technological needs

- developing or refining business models and information system requirements and opportunities and ensuring that the utility's objectives and its customer service goals are well understood by all employees involved in and affected by the new technology

- establishing or updating information system strategies and directions and avoiding the temptation to simply automate current processes

- defining or updating data and applications architecture and ensuring that everyone impacted by the technology has a say in the requirements

- developing or updating technology strategies and a strategic plan that addresses emerging business needs and allowing sufficient time and resources to prepare the information infrastructure that supports the technology, customization, training, performance feedback, and adjustment

Customer Access to Account Information

Because a utility has different customers the utility must adopt and apply different ways for them to obtain information about their accounts. Traditional Monday through Friday office hours do not meet the service needs of many customers, creating a need to assess the way the utility customer service provides accessibility to a diverse customer base.

Most familiar to utility managers are telephone and face-to-face interactions with extended service hours to accommodate customer work schedules. Emerging strategies include online account access, e-mail inquiries, and pay stations located throughout the community (including kiosks in malls or at partnering utilities). Many utilities now offer customers the ability to pay bills online through bank debit programs or credit card payments.

Need for Communicating With Customers

Customers expect the utility to keep them informed about rate changes, capital improvement plans, flushing of lines, new customer programs, water quality matters, maintenance and construction activities, and a host of other issues. The most frequently used methods of contact are billing inserts, direct mailings, print and electronic media, and outreach through neighborhood associations, special interest groups, and schools. Emerging options are public access information, Web sites, and internally produced videos about specific water utility processes and programs that can form the basis for outreach by employees to their constituencies.

There are several essential elements in communicating effectively with customers:

- Honesty. Customers expect truthful answers. The utility's credibility is on the line, as a quick answer that is not accurate is worse than no answer at all.

- The right message to the right people at the right time. Managers ensure that employees are knowledgeable about the subject.

- Established relationships. When the utility needs widespread support for a program or during an emergency, the relationships with those who carry the utility's message—media, community groups, influential leaders and policy makers—are already in place.

CONTINUOUS IMPROVEMENT

Utility managers are challenged to provide levels of service that meet or exceed ever-increasing expectations while maintaining competitive levels of productivity. Key strategies in achieving those objectives include expanding employees' effectiveness with technology; increasing employees' value through training and empowerment; reengineering business processes; managing the load coming from customers; managing customers' expectations by communicating service levels; and identifying opportunities for collaboration, outsourcing, in-sourcing, and enterprising.

Most utility managers believe that they are providing good service based on the anecdotal evidence reported by customer service representatives and in letters (with good and bad comments) from customers. Periodic surveys of customers benefit the utility by helping managers determine how the utility is doing, what it can improve, what new services are desired or expected, and what types of additional information customers wish to have about the utility. This external information is used to help guide a continuous improvement program.

Continuous improvement means a commitment to training and managing a diverse workforce that has divergent goals and needs. Providing high-quality customer service is a demanding job even when the employee has all the necessary tools and knowledge. Rotation and cross-training are excellent ways to improve overall understanding of the utility and to develop a multiskilled work team. An educated and flexible staff solves problems more efficiently, meets peak periods of activity and emergencies better, achieves higher job satisfaction (and, potentially, advancement and compensation opportunities), and allows the manager to exercise greater control over staffing and benefits costs. Managers should always recognize and celebrate successes. Recognizing excellence need not be costly. Recognizing an employee in front of his or her peers is a good motivator. However, recognition is a process that demands consistent application and cannot be perceived as favoritism.

Utility managers should regularly examine business processes for opportunities to streamline them. Process reengineering usually involves modeling the current process (including costs, elapsed time, productivity, and reliability); applying reengineering principles and guidelines; modeling the newly designed process; and testing, evaluating, and fine-tuning it. Those employees who perform the process must be a part of both modeling the old way and helping define the new. Processes reengineered without work team involvement are frequently unsuccessful. Managers must evaluate whether each step in the process adds value and whether the customer is willing to pay for that value.

Outsourcing some customer service functions has worked well for some utilities and has worked poorly for others. In general, managers have greater control over customer interactions if utility staff rather than a contractor conducts them. If customers are the utility's focus, managers must ensure that focus remains sharp and that clear decisions are made. Contract operations companies, energy and telecommunications utilities, banks, and information service companies have positioned themselves to provide outsourcing or collaborative opportunities for utility customer service. To the extent that those collaborative arrangements benefit the utility's customers, they merit evaluation and, where appropriate, implementation.

Customers compare the service they receive from their water utility with that from other utilities and with the range of service providers. A water utility's prime competitors in service and courtesy are found in major commercial enterprises. As the retail industry attests, satisfied customers are less expensive to serve. They also are more accommodating and supportive of the utility's needs to spend what is necessary to attract and keep qualified personnel, maintain infrastructure, and position itself for future requirements.

Excellence in customer service has always been a core value for utility managers, but its impact on the utility's overall success has assumed an even greater importance. Well-trained field and administrative staffs that answer questions, provide information, and evidence a genuine desire to assist customers are among the greatest assets a utility manager can cultivate. Quality customer service is not a by-product of the utility's operations—it is the fundamental reason for the utility's existence.

This page intentionally blank.

Chapter **6**

Operations and Maintenance

OVERVIEW

In operations and maintenance (O&M) management, providing service to customers in a responsible and cost-effective manner is a challenge. Aging infrastructure requires increased budget for either maintenance or construction components. Customer demands for immediate service response impacts staffing and equipment costs and allocations. AWWA has published several manuals dealing with the technical issues of O&M. This chapter focuses on management issues in operations and maintenance.

Utilities are usually organized with O&M functions led by technical managers who are responsible for the actual functions therein. One manager may specialize in water resources, treatment, and pumping, and another manager specializes in maintenance, construction, and equipment operations. While these managers are the experts in their areas, the senior manager must have general knowledge of and be conversant in the requirements and practices of all utility O&M in order to fully integrate the various specializations with other functions in the utility and to effectively oversee budgeting and cost control aspects of the areas.

It is important to know, understand, and use the various tools that assist managers in overseeing O&M functions. Effective and efficient O&M activities enhance the service and worth of the utility. Following is a list of critical tools that should be used by utility managers:

- Records—ranging from work activities and time associated with those activities to testing and repair tasks

- Mapping—accurate locations of system components and easy and immediate access to critical system information by O&M personnel

- Safety—equipment, reporting, and risk reduction

- Training—orientation, skills development, certification, and continuous improvement

- Preventive maintenance—detailed schedules and tasks associated with routine and targeted preventive maintenance on fixed and rolling equipment

- Scheduling—productivity levels for particular work codes; also, evaluation of nonproductive paid time (e.g., travel) to improve competitiveness

- Planning—ongoing assessment of O&M and staffing needs to ensure availability of essential resources at specific junctures in the work process

- Human resources—ongoing training and skills development that results in a multiskilled workforce

- Technology—acquisition of technology, or partnership with others who possess it, to achieve advantages the latest technology offers

- Communications—both improved human interaction and the acquisition of technology that facilitates communication regardless of conditions

- Equipment and tools—regular review of use; establishment of replacement programs; acquisition or lease of specialized equipment used regularly in work tasks to ensure availability of essential resources at specific junctures in the work process

- Policies and procedures—documentation of operating policies and procedures for use in training new employees and in ensuring consistent O&M throughout the utility

- Outsourcing—critical assessment of the utility's core businesses and functions and determination of whether outsourcing provides better ways to allocate current human and material resources

- Specifications and standards—adoption of or upgrading to industry specifications and standards

- Professional associations—tapping into the wealth of knowledge and expertise available through operator and managerial training courses, conferences, seminars, online conferences, etc

- Support services—integrating the work of other utility functions so as to reduce delays, ensure necessary staff and supplies, and maintain essential communications

- Regulatory—integrating regulatory requirements and responses with the utility's business practices, with minimal disruption

RECORD KEEPING

Managers must track work, allocate and use human and material resources, and know when additional resources should be requested through budget items. These tasks are made easier when the utility maintains good records—and then distills and uses that information to improve O&M activities.

Many O&M activities require the expenditure of resources, both for preventive and reactionary work. When preventive or regular maintenance occurs on equipment, such as pipes, valves, meters, fire hydrants, services, points of access, tanks, reservoirs, and pumps, the utility should document that activity. Retained information includes a description of the work done, dates and times, persons and times involved in the activity, tools and equipment used to do the work (including serial numbers, if applicable), and materials required. After this information is evaluated, managers determine the use and costs of resources needed for the work task. The information can also be transferred to a database and aids in determining when infrastructure repair or replacement should occur or whether a particular item has a less-than-desirable length of life. Work history records support continuing improvement decision processes and document information that assists in explaining utility operations and services to policymakers and customers.

Most utilities maintain records of amounts of water treated and pumped, correlating that information to amounts billed or supplied for public health and safety requirements. This information assists managers in determining the need for special leak detection programs or meter replacement programs. Equally important, however, are records of the quantity and quality of work activities performed at pumping and treatment facilities, the level of customer satisfaction, number and types of customer complaints (by category), and such other system information as may enable managers to make informed decisions concerning the types and levels of services offered.

MAPPING

All water systems must locate facilities in order to repair, construct, or relocate lines; provide preventive maintenance; or locate pipelines and services for other utilities to prevent damage when digging. A current set of maps with accurate data is crucial for accomplishing these tasks. Fast-growing utilities are one example of where the infrastructure can outpace the mapping process. Those employees responsible for mapping may need to outsource some of the work to assist the regular staff. This provides the field staff with the needed information in a timely fashion. Many utilities rely on map books with locations of facilities to locate and document repairs and new construction.

As technology has been introduced and become less expensive and more user-friendly, the use of geographic information systems (GIS) has revolutionized the mapping process. GIS locations are placed on base maps as an overlay, providing exact coordinates of latitude and longitude for valves, meters, meter boxes, fire hydrants, pipe segments, storage facilities, etc., resulting in highly sophisticated drawings and related measurements and other data of value to O&M personnel. Utilities that use GIS data are able to print out maps, or workers are able to access both maps and the attendant database information via computers in their vehicles. This improves operating efficiencies by reducing telephone or radio contact and research time. Depending on the level of technology in use, field personnel can prepare *as-built drawings* and complete work orders online, immediately updating maps and infrastructure records and reducing administrative time associated with completing paper documents.

SAFETY

The water utility industry places a high priority on employee safety. Managers must correlate the cost of lost-time injuries with the price of safety equipment and training needed to reduce injuries. A safe working environment is a managerial responsibility. That responsibility includes making sure the knowledge of safe work practices is available through appropriate training; providing equipment and tools that enable safe work performance; conducting periodic work area inspections to ensure that work is performed according to utility standards; and holding supervisory staff accountable for safety matters. All utility employees have an individual and a shared responsibility to

- perform their assigned work in a safe manner, either alone or as part of a work group

- wear appropriate safety equipment and regularly ensure that it is in good working order

- use equipment and tools according to manufacturer instructions, including implementation of safety equipment requirements

- recognize hazards and, where possible, take corrective action to eliminate them

- report safety problems and violations

Personal safety protection for workers includes goggles, earplugs, gloves, steel-toed shoes, hard hats, breathing apparatus, and other protective gear as may be required for a specific job. In addition, tools and equipment, such as harnesses, hoists, shoring, air blowers, and ladders, should be available to provide safe work conditions when climbing or entering confined spaces. Appropriate signs and warnings throughout the workplace denote the presence of hazardous materials, such as chemicals at treatment plants and pump stations.

Worker training is a key factor in utility safety programs and should occur on a minimal basis of once each month with participation a condition of employment. Safety training should discuss the hazards of the job, safe ways to perform duties, equipment to be used, and procedures for reporting unsafe areas or practices. By law (OSHA requirements), accidents must be investigated, safety violations determined and corrected, and unsafe work locations given a high priority for corrective action. Additional safety training on a weekly basis in the form of tailgate-type sessions help remind workers of safety practices in their area.

TRAINING

Well-trained employees who understand what is expected of them when performing their duties are essential in an increasingly competitive environment. Managers must ensure that employees understand the work environment, the organization structure and mission, and their responsibilities when performing work. Knowledge and skills increase over time, and performance management requires managers to identify additional training needs, whether for new work assignments or for improving current skills.

Continuous training is an integral part of good management—some utilities have implemented a training and development requirement of up to 40 hours each year into each employee's performance management plan. Increased competitive pressures and the need to expand employee skills and address pay and benefits expectations have escalated the demand for cross-trained personnel capable of meeting short-term workload pressures or changes in resource availability. State and local regulatory agencies have stepped up the levels of and requirements for certification. Many require certifications before workers are allowed to do certain tasks, conduct certain operations, or attain higher status within the utility. Employees should be encouraged to take advantage of industry professional group, state, and federal training programs for a minimum of 40 hours training per year to reach certification levels and maintain competencies. Managers should assist employees in learning new methods and keeping abreast of modern practices and emerging technologies related to their responsibilities.

PREVENTIVE MAINTENANCE

Utility managers are well aware of the need for preventive maintenance (PM) programs as a primary component in operating and maintaining their utility. However, capital requirements and high costs of equipment often combine to encourage managers to keep costs in check, frequently by repairing facilities and infrastructure many times as *reactive* maintenance occurs. In the short term, it may appear that only reacting to repair needs is inexpensive. However, utilities are, in the long run, businesses, so a preventive maintenance strategy must be adopted and implemented. To the extent possible, all elements of the infrastructure best serve the customer when necessary preventive maintenance results in greater reliability while extending the life of the utility system.

Preventive maintenance includes but is not limited to

- painting pipes in a treatment plant pipe gallery to prevent corrosion

- testing valve and fire hydrant operations on a defined schedule to give reasonable certainty that they work properly when needed

- scheduling preventive maintenance—lubrication, alignment, and heat testing of bearings—on pumps and motors

- conducting inspections and coatings of storage facilities on a scheduled basis

- conducting cathodic protection and leak surveys to ensure safe infrastructure and minimal water losses.

The importance of integrating a comprehensive PM program cannot be overstated. Best-in-class utilities that insist on providing reliable services to their customers implement such programs and closely monitor their results as a component in determining both annual and capital budget requirements. The amount

Figure 6-1   Preventive maintenance allows a utility to have a long-term future

of PM varies from utility to utility. However, the accepted rule of thumb is that PM encompasses about 80 percent–90 percent of the work effort, with reactive maintenance being 10 percent–20 percent of the effort. It is highly unlikely that a utility will never have to react to a problem in the utility's infrastructure. It *is* highly likely that a utility that adopts an aggressive PM program will have a lower amount of resources allocated to reaction and fewer instances of customer inconvenience.

Regular reviews of the preventive maintenance program by managers within the organization and analysis of customer comments will ensure a higher level of success. This identifies work process improvements along with needed staff reorganization.

SCHEDULING

Personnel costs are a large part of the utility's budget; it is imperative that managers use their available resources in the most effective and efficient manner. Utility managers should periodically monitor work processes to determine whether the allocation of human and equipment resources meets the challenge of high-quality performance and optimal service. If scheduling is done correctly, the appropriate equipment, tools, materials, and personnel are on-site to accomplish work as safely and as quickly as possible, keeping costs low while meeting service expectations.

Operations and maintenance activities usually vary among utilities, and it is impossible to definitively identify the appropriate size of staff and work teams and times for work shifts. Regular evaluation of productivity—possible through good data analysis from work orders or other records—helps to determine the proper size of work teams and the approximate time required for the scheduled work effort. The availability of cross-trained personnel enables a manager to gain the greatest flexibility in staffing routine and emergency work projects. One way to facilitate greater effectiveness and efficiency is to have work projects identified and scheduled when employees report for work. In many instances, this requires the manager to plan and schedule the next day's work activities at the end of the prior shift. The advent of technology—wherein an initial work assignment is scheduled, with additional work orders transmitted to computer terminals in work vehicles—is an asset in increasing effectiveness and efficiency. Rather than completing work orders in paper format and having additional staff enter that data into the records management system, data are compiled in the field while the work team is moving to its next assignment. Additionally, field drawings can be entered and measurements checked at the site. Obviously, work schedules change if emergencies or unanticipated repairs take priority.

Optimal use of trucks, backhoes, trailers, and specialized equipment requires that managers continuously know work team needs and equipment availability in order to deploy resources where they are needed. Equipment is shared in utilities with multiple work shifts. If work teams have specialized assignments, specific equipment may be assigned to specific work teams.

PLANNING

Even more critical to O&M success is improved managerial planning for the activities, construction, repairs, and replacements of the utility's infrastructure. Enhanced coordination within the utility and with agencies outside the utility is an essential component when improvements impact other functions. For example, when a water main segment is to be taken out of service, O&M managers must coordinate those activities with customer service, production and public information staff, as well as with other utilities, public safety personnel, and the customers who will be affected.

O&M managers have a critical role in the operating and capital improvement plan (CIP) budget processes. The budget process is no small task and is absolutely essential to successful O&M. As pressures on budgets increase—doing the same or more with the same or less—managers have a greater responsibility for determining program and project needs based on well-documented data about the activities being managed. Defining the utility's needs for operations thus involves reviewing past expenditures, activity levels, age of infrastructure and equipment, and human resource requirements.

Capital improvement budgets address major system needs, including new facilities to accommodate growth and major replacements to improve the quality of service. Major replacement projects included in the CIP process require good documentation. This information is distilled from work orders and a variety of other sources, such as the number of breaks to a pipeline, number of failures of a motor, the condition of water storage facilities, and indications of inaccurate measurement of water use.

HUMAN RESOURCES

Attracting, improving, and retaining good employees is a challenge to any manager, particularly in a competitive employment environment. The manager must not only be aware of the knowledge, skills, and abilities required for each position within the utility but also of local labor market conditions. Managers are also advocates for compensation and benefits that reflect the importance of the public health and environmental quality responsibilities entrusted to utility employees. The chapter on Human Resources provides detailed information on human resource requirements.

It is important that managers ensure that employees understand the mission and goals of the utility and how employee work assignments fit within the greater organization. At the frontline level, managers and their employees should review the position descriptions to ensure that work activities are accurately described and that the knowledge, skills, and abilities reflect those needed for optimal job performance. Regular performance reviews are a key element in good work management and should not only inform the employee of how well work is being performed but also identify areas for improvement and individual development possibilities that enable the employee to expand his or her skills and earning capacities.

Managers are responsible for establishing the quality and health of the work environment. Ideally, the work environment is one in which each employee feels valued and a part of a team that solves problems, offers innovation, provides good service, and gets the job done efficiently. Managers are also key individuals in the utility who provide role models for employees and encourage self-development and career considerations.

Many utilities strive to create a career environment where employees develop their knowledge, skills, and abilities and move to positions of increasing responsibility and reward. Utility managers must also consider the need for external hiring when necessary, in order to add new skills and to challenge the growth of both the utility's mission and its staff. External hires may bring new ideas and methods that improve work processes, reduce costs, and better apply scarce resources.

Manager and employee work teams can assist one another in developing skills through cross-training while also improving the utility's ability to meet its service objectives. Frequently, raising employee skill levels allows for improvements in position classifications and compensation. Thus, operators at water plants may perform maintenance activities, such as adjustments of chemical feed equipment, lubrication of equipment, and diagnostics on process equipment. Field work teams can cross-train in valve, pipe, fire hydrant, and meter repair. Management gains greater flexibility in assigning work activities, and more complex maintenance activities can be assigned to technicians with different skills. Overall compensation levels can be adjusted to reflect the breadth of skills required to complete multitask work assignments.

There is an emerging managerial focus on succession planning. Long-term employees complete their service to the utility and retire, in many instances at the same time that competitive pressures in the labor market impact the length of time that newer employees remain on the job. Managers are challenged to ensure that employees at every level in the organization have an opportunity to prepare for the future and to anticipate the need for fully qualified personnel to perform higher-level work, whether temporarily or in anticipation of future position vacancies.

TECHNOLOGY

Strides in technology have led to major impacts on utility business functions. Computers—long found in administrative and customer service venues—are now found in field operations, work order management, and data acquisition concerning production and distribution activities. Networking has made it possible to share information across the organization and with other entities. Consumer applications, such as Web-based ordering and "shopping carts," have been implemented to coordinate inventory management with accurate and rapid restocking of maintenance and construction trucks.

GIS data assists in locating infrastructure in field settings; work order information and utility maps are also available on field computers.

Computer maintenance management systems (CMMS) enable development of records related to work effort, materials, equipment, tools, and staffing. Entries for work to be done, from its inception to its completion, are easily tracked. This enables managers to make adjustments to work processes and assignments or to set improvement goals where deficiencies are apparent.

Supervisory control and data acquisition (SCADA) systems allow the use of technology to assist in controlling treatment processes, monitoring system component performance, and recording hours of equipment use. Management reports are configured as needed, and the obtained data helps meet regulatory requirements.

Water meters are still manually read in many utilities, with some using hand-held computers to record readings while others use a paper-centered recording process. However, the need to control costs, coupled with the rapid deployment of technology at a cost-effective level, has led to widespread use of automated meter reading systems—from touch-read systems to drive-by signal readings, to telephone dial-up readers. The objectives are more accurate meter readings, reduced risk of accidents and injuries, and improved cost controls through resource allocations. For more information, see AWWA Manual M6, *Water Meters—Selection, Installation, Testing, and Maintenance*.

COMMUNICATIONS

Organizations that provide goods and services to customers through various processes require teamwork by organization workers. Teamwork requires use of a good communication system. The communications chapter in this manual focuses on communications requirements in greater detail. For purposes of this chapter, however, it is critical to note that every manager and every supervisor is a communicator who shares information through all levels of the organization and solicits and receives feedback. This means sending information to senior managers, receiving and transmitting information to and from employees, and collaboratively communicating across functional levels to expedite work programs and activities. Too often, information becomes a hallmark of power and is retained rather than shared. This occurs most frequently in parts of the organization not considered to be knowledge-based. As a result, subordinates are not conversant with the utility's vision, mission, goals, and means of achieving them—they are not in the loop.

Managers are responsible for informing their employees through meetings, written communications, or other available methods. Changes in processes, procedures, and plans are essential communications topics. Performance of the work group, division, department, and utility as a whole should be updated at least every quarter. Two-way communications provide the best opportunity for dialogue and greater understanding. Reports to and from senior management must focus on organizationwide activities and interdependencies. Reports should include all information needed to support work efforts of the utility, as well as answer questions directed by both superiors and subordinates. Monthly, quarterly, and annual reports provide opportunities to track and provide information to both customers and employees.

EQUIPMENT AND TOOLS

Managers are responsible for knowing the equipment needs of their work teams and for budgeting and acquiring the necessary tools and equipment to perform assigned work. Tools and equipment must be kept in good repair and inspected for safety as a part of routine O&M. Work teams should expend as little time as possible locating the tools and equipment they need to perform assigned work tasks. Managerial control includes

- periodic inspections of tools and equipment to ensure that work teams have what they need and comply with their responsibilities for its care

- regular monitoring of work project records to determine whether multiple trips are required to obtain the materials, equipment, or tools necessary for the job

To the extent possible, managers should have a standard set of tools and equipment assigned to each work team and their vehicle, with the work team accountable for breakage, loss, or improper use. Work teams are responsible for maintaining assigned equipment, tools, and vehicles. Routine vehicle maintenance should be scheduled so that work teams are not idle. Breakdowns and repairs should be addressed immediately. Equipment services are ideally performed during nonshift hours so that vehicles, backhoes, and other tools are ready for use at the beginning of the workday.

Equipment reliability requires keeping maintenance and repair records, monitoring performance problems, and establishing a replacement schedule. Based on documentation and the utility's general replacement schedules, managers should budget funds to acquire equipment as a matter of continuing improvement, rather than having an inordinate number of equipment replacements in any one year.

POLICIES AND PROCEDURES

O&M managers are responsible for developing and implementing policies and procedures directly related to the functions they oversee and are accountable for implementing and complying with administrative, financial, and other policies and procedures that govern the workplace.

Examples of typical operating policies and procedures are safety requirements, opening and closing trenches, PM, use and care of tools and equipment, record-keeping, reporting accidents, security, and vehicle use and care. Specific policies and procedures might also include, but are not limited to, work task or treatment protocols and emergency notification.

Fair and equal treatment of employees is an essential component of management, and written expectations and consequences for inappropriate behavior are necessary to provide direction. Employees must understand the terms and conditions of their association with the utility, including work performance, attendance, behavior, regulatory requirements, safety, and interpersonal relationships. Written job duties and processes, as well as how performance is evaluated, should be shared with the employee. Other written materials should outline employee rights, benefits, promotional expectations, and recourse for specific problems encountered in the workplace. If employees fail to meet the established expectations, consequences in accordance with disciplinary procedures may be necessary. The manager must ensure that appropriate documentation is maintained.

Administrative and financial policies usually apply to the entire organization in terms of topic and implementation. It is important that employees know, understand, and comply with policies concerning ethics (especially in terms of receiving gifts from vendors, contractors, suppliers, etc.), standards of conduct in dealing with the public, conflict of interest, expenditure of utility funds, personal use of utility equipment, and unauthorized communications.

OUTSOURCING

O&M managers must recognize their responsibility in ensuring the utility's competitive posture. This may require an evaluation of business processes and, ultimately, determination of whether some work must be outsourced. Activities that are not a primary responsibility—a core function—of the work group should be identified and the benefits of outsourcing assessed. As an example, some utilities have mechanics or operating personnel perform groundskeeping functions at plants or pumping stations. This work assignment is secondary to their primary task of maintaining pumps and motors. If it is deemed by managers a reasonable use of employees' time, it may be continued. However, managers must determine whether outsourcing the secondary work tasks improves mechanical maintenance or reduces costs. Other potential areas for business process reviews and potential outsourcing include highly sophisticated laboratory analyses where equipment costs and utilization cannot be justified; some engineering, construction, and specialized equipment requirements; and motor repairs.

SPECIFICATIONS AND STANDARDS

O&M personnel have a critical role in the early design of infrastructure, especially when determining and establishing specifications and standards in design. Those who operate and maintain water distribution systems often know where weaknesses in design and standards affect performance levels they are charged with maintaining.

Early input into the process of setting specifications and standards is highly valuable for eliminating problems that are obvious to O&M personnel.

For parts and materials needed by O&M work teams, there is merit to the argument made for standardization. Selection of a limited number of types of materials or appurtenances means that fewer repair components must be stocked, less warehouse space is needed, and savings in inventory occur.

The utility must have written specifications and standards for its systems, especially when others (such as contractors or developers) install infrastructure that the utility must maintain. Many utilities adopt and specify AWWA Standards for infrastructure components. There are a variety of industry- and manufacturer-accepted components that can be incorporated in the utility's project approval process. Inspections by utility representatives of new or replacement projects ensure that the correct materials are used and that they are in compliance with approved project plans. Allowing substandard materials and installations means the manager has accepted a maintenance problem for years to come.

PROFESSIONAL ASSOCIATIONS

As in any industry, there is a lot of information to be gleaned from peers. Managers should be involved as active members in local, state, and national and international organizations that focus on improving water utility operations, maintenance, and management through sharing information and technologies. Organizations such as the American Water Works Association (AWWA) provide forums for training and education as well as a library of industry-related materials and research. In addition, many state and regional organizations promote professional development, operator certification, and environmental excellence. While this area is often subject to budget scrutiny, it is a key investment that the utility makes in ensuring that its staff is knowledgeable and is using or planning for state-of-the-industry improvements.

SUPPORT SERVICES

Managers rely on support services to expedite operations and maintenance activities. Support services may be assigned to an administrative function manager rather than being a direct reporting unit of O&M. Regardless, the manager must establish the lines of communication and accountability necessary to ensure that the support services are available and easily accessible. Support services can include but are not limited to

- Warehouse and Inventory Management. Parts, tools, materials, and other requirements for the work task must be available in sufficient quantity and reordered in a timely manner so that work activities are not held up. Most utilities have an inventory management system in which the work order records of items used are compared with warehouse stock and new items are ordered when a predetermined inventory level is reached. Many utilities have adapted the *just-in-time* approach for maintaining stores—delivery of materials, parts, etc. (particularly high-cost items), occurs when they are needed instead of having utility revenues tied up in inventory. The *shopping cart* concept is also another way to ensure that field inventories are fully supplied and work teams do not lose critical repair or construction time because they lack a particular item. Finally, managers must ensure that work teams have access to essential materials and supplies for after-hours emergencies.

- Security. Utility managers are more aware of the potential for internal and external threats—from physical violence to cyberterrorism. As a result, there is greater emphasis on establishing security controls and on monitoring use of and access to physical facilities and information systems. As part of their mandatory vulnerability assessments, utilities have increased the level of security, staff capacity to respond to emergencies, and general access to information (including that located on Web pages or in publications). Managers need to ensure that security professionals, whether utility employees or contracted, are familiar with all O&M activities, locations, potential dangers, and personnel identifiers so they can perform their assigned tasks.

- Dispatch Services. All utilities rely on a central emergency number for notification about service problems, main breaks, and other emergencies. Dispatch personnel are usually utility employees, but may—especially during shift hours—be from other departments or from contract service providers. Managers must be sure that all dispatch personnel are familiar with emergency response requirements and that they have an up-to-date listing of key personnel to be notified (including not only utility personnel, but also local public safety agencies). Dispatch operators need to have sufficient orientation to the O&M work processes so that they obtain the correct information concerning system problems, identify the correct locations and other critical data, and give clear instructions to both customers and to field personnel who respond to the situation. They must also have sufficient training to monitor SCADA equipment and other system-sensing devices, make immediate operating decisions, and contact the correct utility representatives to respond.

Accounting, financial, and procurement activities are usually assigned to specialized service groups. Customer service groups also interact regularly with O&M groups. They must be kept informed of major emergencies in the system or of planned maintenance and construction so that they, in turn, inform customers who seek information about service problems. The utility may have or may contract with other work units to repair streets and sidewalks when projects are completed.

It is worth noting that support service groups frequently are cited as the reason work activities are held up or why work teams and individuals are not able to do their job. In reality, support service groups provide extensive assistance in expediting O&M work tasks if there is ongoing communications and a common goal to serve the customer. There is no question that work processes in different areas can constrain O&M personnel, leaving managers with the belief that their efforts are delayed by red tape. However, managers have a responsibility to establish and maintain communications with support services groups, clearly explain their needs, and seek a solution that solves any problems and strengthens working relationships across functional lines.

REGULATORY

Regulatory requirements and oversight are a continuing challenge for utility managers. Whether through legislation or by agency rule-making, regulatory personnel, and the agencies they represent, have job requirements that emanate from public concerns and opinion about health, safety, and environmental issues. Those requirements grow out of public concerns and public opinion about health, safety, and environmental issues. Many regulatory requirements have improved operations, safety, and health and environmental quality. Training has added to the professional capacity of employees, especially when certification standards are met and training completed. Other regulations relate to how work is performed while protecting the health and safety of customers and workers. Managers must monitor and document regulatory compliance to communicate the safety of utility water to consumers. Regulatory agencies can be an important ally in achieving that objective. Professional associations, such as AWWA, provide help in understanding and complying with regulations, as do local sources, such as other utilities, consultants, and the regulatory agency itself.

Larger utilities may create their own internal regulatory group for self-monitoring. This is especially true with state and federal budget crisis situations. An internal regulatory group can guide the entire organization through the compliance maze and assist with communication between state and federal regulators and the utility.

O&M consumes a significant portion of a manager's time, efforts, and resources. Successful managers are well-versed in the components, but utilize subordinate managers and supervisors who have the authority to oversee the utility's programs, are held accountable for their actions, and are responsible for overall O&M success.

This page intentionally blank.

Chapter **7**

Environmental Health and Safety

OVERVIEW

The water utility industry is charged with providing for public health (drinking water) and environmental quality (wastewater treatment, watershed management). To meet those obligations, utility managers must be knowledgeable about and fully compliant with a variety of national, state or provincial, and local laws and regulations. Some states and provinces may establish higher standards than those set at the federal level. More important, because of the nature of the work utilities perform, they must establish strong health and safety programs to protect employees and the general public. This chapter reviews some of the key points managers must address in these areas. Compliance is less costly than accidents, injuries, and fines and causes less stress on the organization and its resources.

ENVIRONMENTAL LEGISLATION

Utilities must comply with specific provisions of several environmental legislative initiatives to protect air and water, to address resource conservation and recovery matters, and to ensure the viability of the utility in emergency situations while protecting the public's right to know about a particular event. It is important to remember that while compliance has been mandated for many laws and their attendant rules, most have resulted from public concerns about the laws' topical issues. Honest efforts in compliance and reporting requirements are key to successful implementation and build credibility within the community as the utility tells its story and explains the associated costs and programs.

The principal US laws that affect water utilities are the Clean Air Act, the Clean Water Act, the Resource Conservation and Recovery Act, the Safe Drinking Water Act, and the Emergency Planning and Community Right-to-Know Act. Other federal and state legislative acts may have articles that affect utility operations. The best information about these and other state legislative initiatives come from two sources:

- The Internet allows managers to search for specific information about an act by entering the appropriate initials. A variety of sources are available, but the most definitive is www.epa.gov, which has drop-down menus to each of the environmental areas it oversees. Many US Environmental Protection Agency regions and state agencies maintain Web sites that are accessible for more direct local data.

- Regulatory agency personnel, especially state and provincial employees who perform required site inspection surveys, are knowledgeable about many regulatory requirements and most are accessible and willing to assist utility managers with information and compliance.

Water supply in Canada is managed at the provincial level, thus there is no national jurisdiction unless federal institutions or first nations are served. Several provincial governments have laws and regulations to address their specific requirements, including environmental issues. It is important to ensure sustainability of the water system's assets. AWWA's Canadian Affairs Council provides additional information. In Mexico, environmental issues are addressed by both federal and state agencies. It is noted that the United States, Canada, and Mexico have many joint programs and projects that address environmental concerns of all nations in North America.

Utility managers are responsible for their organizations' compliance with provisions of all environmental laws that affect their business. Thus, it is important that a utility manager has a solid understanding as to requirements within the law and that he or she assigns authority for ensuring compliance to the appropriate subordinate area. Because the manager holds ultimate responsibility, it is imperative that he or she be kept abreast of the utility's position and needs so that funding or program priorities are adjusted to meet compliance requirements.

RISK MANAGEMENT AND INSURANCE

Each utility, as part of its basic business practices, maintains insurance to cover its property, personnel, and emergencies that might occur. Generally, this insurance coverage is maintained through an external provider who handles claims and who may provide training, voluntary inspections, or other services. Typically, utilities maintain liability, replacement, damage, and vehicle insurance through commercial carriers. Each type of insurance has some degree of regulation through state or provincial government insurance commissions, which assure a modicum of consistency as to practices, coverage, rates, and protocols for claims. Utilities normally insure with commercial carriers who have experience in utility coverage and who assist the organization in determining the correct deductibles, ranges of benefits, etc. A good working relationship with a carrier's agent is a strong asset for a utility manager and the risk management staff.

Some utilities prefer to self-insure, and this practice is frequently chosen for employee benefits insurance—health, life, disability, retirement, catastrophic, etc. The utility then hires a benefits administrator to establish payment rates for services, approve procedures, and process claims. It is important that the utility manager and risk manager know the requisite state procedures for self-insured utilities and that those programs are continually monitored for fiscal accuracy and desired service levels.

Workers' compensation insurance is required of each utility, whether it is self-insured or covered under a state plan. Utility and risk managers must be highly

Figure 7-1   Water utilities are at risk and must carry insurance

knowledgeable about the time elements, reporting requirements, and rules and regulations of workers' compensation programs, as they differ from state to state. Workers' compensation systems are a significant resource for the utility in terms of reducing costs and identifying and correcting work procedures, equipment, or equipment usage that frequently results in on-the-job accidents and worker injury. In addition:

• Most workers' compensation agencies provide training and programs for risk managers and, in some cases, for all utility personnel. These training programs include materials and videos that address worker injury matters.

• Many agencies also provide consultation services to utilities, in many cases conducting on-site inspections or audits.

• Resources suggested by the agency that help ensure compliance with other health and safety requirements can, if implemented, result in reduced premium and payment costs.

PUBLIC NOTIFICATION

The expectation of a wide range of public notifications concerning utility problems and operations has grown in recent years. As a result, many utilities are faced with many circumstances where notification is expected and required. Those circumstances can include

• contamination

• security breaches and plant break-ins

• nitrates

• security alerts (homeland security issues)

• arsenic

• boil-water alerts

• main line breaks

• high-water alerts

• traffic pattern changes

• power outages

In addition, citizens are increasingly interested in knowing about utility work zone areas—what is to be done and the timetable for repairs—and weather events that may affect distribution or water quality (tornadoes, major storms, flooding, and hurricanes). Also, the public must be notified about chemical releases or spills, regardless of whether they occur in a contained area or whether they are in public property.

When an event occurs that requires public notification, the utility is usually busy exercising its technical capacity to respond to the issue—leaving little or no time to develop a written document describing the situation. It is in the manager's best interest to develop a template that contains the required language and leaves blank spaces where the critical, incident-specific information is entered. This facilitates preparing the notifications as an administrative process rather than having to assemble technical or subject-matter experts to draft the language when their expertise is sorely needed elsewhere. The language that is necessary for the template—in addition to that which is required—includes what the notification is about, who is affected, what the water supply can be used for (e.g., sanitation), and where additional information can be obtained (contact names and telephone numbers). After notification is given, concerned citizens are likely to call the utility, and customer service staff must have essential information about the notification in order to respond professionally to inquiries.

When public notification is required, utility managers are likely to be in contact with the media, including radio, television, and newspapers, but there are other tools available to fulfill notification requirements. The utility's Web site can carry a copy of the notification and, by having it prepared, utility personnel can quickly duplicate flyers for door-to-door distribution to customers or for public posting. Other reporting should include the utility's emergency management team, the health department, police and fire agencies (city, county, and state or provincial), and other outside agencies that might assist in the utility's response or require special notification, such as the Red Cross or the Department of Transportation. An example of a boil-water order is found in Figure 7-2.

In serious emergencies, managers must ensure that they have access to contractors, vendors, consultants, laboratories, equipment rental agencies, and emergency equipment contractors for unique requirements. The requirements include snow or debris removal, environmental cleanup, health and wellness follow-up, and special equipment, such as cranes, derricks, large haulers, and related equipment, e.g., water-haulers.

Utility emergency response plans should include policy, protocol, and procedures and are required to be in writing and accessible to and understandable by all employees. When public notification is required, utility staff cannot be mired in locating the adopted procedures and protocols. They need to be dealing with the situation. This part of the process is the responsibility of managers and supervisors who must ensure that all staff members know their responsibilities and that they will be held accountable for complying with those requirements.

One of the most important tasks of a utility manager is to conduct practice drills of protocols before an actual emergency occurs. Employees should be trained on utility procedures, governing policies and protocols, and other areas as appropriate to ensure the desired response level. Training topics include personal protective equipment, handling difficult people, working under stressful conditions, workplace violence, and the utility's basic commitment to quality service.

Maricopa County
News Release

Environmental Services
Department
1001 N. Central Avenue
Suite 550
Phoenix, AZ 85004
Phone: 602-506-6611
Fax: 602-506-1874
www.maricopa.gov/envsvc

For additional information:

Johnny Diloné, MCESD Public Information Officer
602-506-6611, Cell Phone: 602-525-2423
Emily Poland, MCDPH Public Information Specialist
602-506-6607, Cell Phone: 602-722-1806

January 25, 2005

BOIL WATER ADVISORY:
SAFETY MEASURES REQUIRED

Maricopa County Departments of Environmental Services and Public Health advise all water customers served by the City of Phoenix to boil water used for consumption and take necessary precautions until further notice. The Boil Water Advisory was issued this morning by City of Phoenix officials due to the detection of high sediment levels, known as turbidity, found to be entering the water supply at some City of Phoenix water treatment plants.

This precautionary health advisory is extended to all City of Phoenix water customers, including residents in Phoenix, Tolleson and other areas served by the City of Phoenix water supply.

Be advised that water needs to be at a boil for at least 5 minutes before being used in the following situations:
- Drinking water
- Washing dishes
- Brushing teeth
- Food preparation
- Making ice
- Wound care

The Boil Water Advisory affects the health and safety of all eating and drinking establishments including retail groceries operations, restaurants, schools, daycare centers, hospitals, senior care facilities, etc. Food establishments need to follow special standards to remain in operation. Please visit the following link for a list of the required items:

www.maricopa.gov/envsvc

Schools and day care centers should prevent children and staff from consuming water from water fountains. If bottled water is unavailable, necessary actions should be taken to prevent children from dehydrating, particularly after any physical activities.

-more-

Courtesy of Maricopa County, Arizona

Figure 7-2 A sample boil-water order issued by Maricopa County, Ariz.

Maricopa County NewsRelease
Add Descriptive Text Here
Page 2 of 2

"If an establishment is not capable of meeting these health and safety standards, they must remain closed until the Boiled Water Advisory is lifted by Maricopa County Environmental Services," said Al Brown, Maricopa County Environmental Services Director.

Additionally, residents affected by this current water safety situation are encouraged to follow a water conservation plan as recommended by City of Phoenix and Maricopa County officials. This may include limiting bathing or showering, washing cars, and watering lawns and plants until further notice.

High sediment or turbidity levels interfere with the disinfection process and provide a growing medium for microbial growth.

"Turbidity may indicate the presence of disease causing organisms, but it is important to know that water may still look fairly normal," Brown said. "Therefore it is essential that residents follow proper water and sanitation precautions, until notified by County Environmental Health officials that the safety of the water has been assured."

"This has the potential to be a serious public health issue, but we're taking all necessary precautions to prevent disease in the county," said Dr. Doug Campos-Outcalt, Medical Director for the Maricopa County Department of Public Health. "The role of the Department of Public Health in such situations is to advise residents of potential threats to their health, to advise them of appropriate precautions to take to protect their health and safety, and to communicate to them when the health threat is over."

The MCDPH is working with the community and is notifying health care providers and facilities to watch for patients with gastrointestinal disease symptoms. Residents who may experience nausea, vomiting, diarrhea, and abdominal pain should consult a health care provider as disease symptoms could occur from 1 to 12 days after disease exposure.

Please be aware that this Boil Water Advisory will be in effect until lifted by the Maricopa County Environmental Services Department.

For more information private citizens may call the City of Phoenix Water Department customer service line: (602) 262-6251.

The Arizona Department of Environmental Quality has the following link for emergency disinfection of drinking water: http://water.azdeq.gov/envion/water/dw/download/desinfec.pdf

###

Figure 7-3 A sample boil-water order issued by Maricopa County, Ariz. (continued)

OCCUPATIONAL HEALTH AND SAFETY

Utility managers must demonstrate leadership and excellence in all aspects of operations. At a minimum, they must strive to operate their organizations in compliance with environmental, health, and safety laws and regulations. More important, managers are responsible for ensuring a safe, healthy, and productive workplace through training, compliance programs, and management systems that identify and address risk factors. Managers play a key role in

- developing and administering industrial hygiene and safety

- defining compliance requirements for each of their operations

- supplying guidance to supervisors and employees in implementing the adopted health and safety program

- regularly measuring their organization's progress, making corrections to improve conditions or processes, and ensuring that staff compliance with health and safety protocols are a measurable part of performance management

This type of leadership enables overall guidance to and oversight of the environmental compliance program and establishes a framework for developing programs, policies, and procedures that support the utility's efforts to comply. Most utilities are either required to comply with the provisions of applicable laws and regulations or they choose to do so because it makes good business and economic sense. Utility operations often include many processes that are inherently dangerous. It is imperative that managers place strong emphasis on an aggressive program that includes the following:

- Engineering—substituting less hazardous materials, reducing inventory, modifying procedures, designing out hazards, incorporating fail-safe devices, using warning devices, and prescribing personal protective equipment.

- Education—training personnel in safe procedures and practices, teaching staff how to do a job safely and correctly, and teaching users how to safely use a product. It also includes identifying and training staff about hazards that exist in a product, process, or task and how to take appropriate protective actions.

- Enforcement—achieving compliance with all laws and regulations, with consensus standards, and with utility rules and procedures.

A good health and safety program includes attention to controlling hazards. This means the utility

- requires the use of personal protective equipment

- regularly inspects tools and equipment for safety compliance

- regularly inspects work sites to determine whether work teams are complying with health and safety rules

A safe work environment does not happen by itself—it can only exist when there is a joint commitment of managers and employees to establish and maintain a safe working environment, and it exists when safety is recognized as an ongoing commitment, not a one-time response to an adverse event. No organization wants to see a worker buried in a trench, inhaling toxic chemicals, or involved in an equipment or vehicle accident, and no employee wants to be in that position. Yet, accidents and injuries continue to happen. In simple terms, safety is a matter of personal accountability.

Each employee is responsible for complying with utility environmental, health, and safety rules and with applicable standards mandated by federal, state or provincial, and local laws. Employees must understand that seeking and accepting responsibility means taking initiative and ownership and being accountable for actions. A good place to start emphasizing personal responsibility for safety is in both the position description and in on-the-job orientation and training. The following must be an often-repeated philosophy:

> *Every utility employee is responsible for working safely—alone or as part of a work group—and for wearing the appropriate personal protective equipment and maintaining tools and equipment at the highest possible level of performance or for obtaining replacement items as quickly as possible. Every utility employee is also responsible for identifying and reporting safety hazards and, where possible, correcting them at the time they are identified.*

This level of commitment should appear as part of every job description within the organization, and the philosophy must be found throughout the organization if accidents and their costs are to be reduced.

Management support for safety programs and safety training is essential. Managers must establish and manage safety and training programs as assets for the utility. That means identifying and financing resources, periodically auditing program offerings, and insisting on compliance with rules and regulations. Utility managers should develop safety manuals and provide easy access to these manuals. This includes a mission policy statement, assigned responsibilities, manufacturers' safety data sheets, and a control mechanism that ensures the overall safety program meets its objectives. Most utilities assign safety responsibilities to an individual who is then accountable for investigating accidents and injuries, maintaining records, and reporting. The latter two components are essential—if records are not maintained and analyzed, no followup or corrective action takes place.

TRAINING

A principal objective of safety training programs is to prevent accidents and injuries. All good safety programs begin with a needs assessment that identifies and analyzes job hazards. It is important that the individual responsible for safety works with managers, supervisors, and employees to determine organizational and individual safety training needs. Each utility must develop the appropriate content and delivery mechanisms and on-the-job-training, deliver and evaluate the training efforts, determine what type of followup training is required, and determine when, where, and how frequently it will be offered. Each organization must also monitor overall progress and effectiveness of its programs. Some utilities measure and prove employees' knowledge by quizzing or testing their retention of the safety training that was provided.

Safety training should encompass all hazards encountered on the job. These may include, but are not limited to, chemical, electrical, work site, vehicle, confined spaces, asbestos pipe, fire, material handling, and excavation dangers. Training should also equip personnel to perform work-site safety analyses. These courses are the most successful when taught by qualified and competent trainers. It is equally important to have trainers available on a regular basis so that all personnel on all shifts are able to attend and learn.

It is to the utility's benefit to ensure that enough employees are familiar with basic first aid techniques, cardiopulmonary resuscitation, and automatic external defibrillators to immediately respond to emergencies. From a safety and security standpoint, the utility may wish to obtain and install an emergency alarm for individuals who work alone, on shifts, or in areas where high levels of violence or crime are reported. Table 7-1 identifies other programs that should be regularly offered to staff.

All utilities must maintain training record documentation that meets reporting .requirements and continuing workers development standards.

Table 7-1   Safety training programs

Training	Program Certification
• Asbestos	• Lockout/tagout
• Confined space	• Process safety management/USEPA risk management plan
• Back belts	• Confined space
• Bloodborne pathogens	• Trenching and shoring
• Electrical safety	• Hazards communication
• Forklift	
• Hazards communication	**Inspections/Maintenance**
• Hearing conservation	• Cranes/hoist
• Lead	• Eye wash stations
• Lockout/tagout	• Drenching showers
• Emergency preparedness	• Elevators
• Excavation, trenching, and shoring	• Emergency lighting and exit signs
• Fire extinguisher	• Fire extinguishers
• Hazardous waste operation and emergency response	• Fork lifts
• Powered hand tools and machinery	• Respiratory equipment
• Personal protective equipment	• Slings
• Process safety management of highly hazardous chemicals	• Welding/cutting/brazing
• Job safety analysis	• Chlorine repair kits
• Fall protection	• Pumps
• Respiratory protection	• Trailers

RECORD KEEPING

An essential element in an effective health and safety program is the management of injury or illness records, as well as medical surveillance, exposure monitoring, and accident data. This information provides the basis for meeting regulatory reporting and record-keeping requirements. It is also a valuable tool for tracking workplace trends, evaluating the effectiveness of the utility's safety and health program, and defining new directions for improvement.

In some instances, government agencies require immediate notification if an employee is killed or more than one employee is hospitalized as a result of the same work-related incident. There are specific notification processes that may include state or provincial and federal agencies. Regardless of whether such notification is mandated, all supervisors should be required to document the following information:

• facility name

- location of incident

- time of incident

- number of fatalities or hospitalizations

- contact person

- phone number

- brief description of the incident

Additionally, the utility—and, potentially, state or provincial agencies—should maintain incident reports on both vehicular and work-related accidents involving utility facilities and personnel.

INCENTIVE PROGRAMS

A key tool in the utility manager's kit is providing recognition for safe work behaviors and for completion of required safety training. Whether it is a safety awards ceremony, recognizing a crew for responding to an emergency, or an annual banquet that focuses on safe execution of the utility's work projects, recognition is appreciated and motivational. Managers can publicize awards, arrange for interviews by local reporters with utility employees who successfully dealt with an emergency, or simply make notice of staff on official bulletin boards or in newsletters. The utility should maintain training record documentation that meets requirements of state or provincial and federal agencies for particular types of continuing worker development. Incentive programs must not encourage underreporting of accidents and injuries by employees or supervisors.

A successful safety program requires continual followup on the progress of training to determine whether accidents and injuries have increased or decreased. AWWA's safety committee, in conjunction with national organizations, recommends maintaining a five-year accident/injury trend line that helps managers evaluate current safety policies, programs and training. Because accidents and injuries result in significant costs to the utility, managers need to develop cost analysis figures and trends for accidents and injuries. Far too often, employees do not realize the extent of these costs, and many utilities are now sharing that information as part of an effort to improve performance.

Many resources are available to utility managers through AWWA and other vendors, consultants, utilities, and state emergency management agencies. Managers must use them in a conscientious manner with the understanding that safety is not just a cost, it is a sound business practice that is worth the utility's investment.

Workplace safety is smart business, and managers play a critical role in ensuring that all resources—staff, equipment, processes—contribute to an efficient and effective level of service

AWWA MANUAL | M5

Chapter **8**

Security

OVERVIEW

Security has long been a concern for water utility managers and employees. However, the prospect of water utilities as targets of terrorist acts has forced managers to place more emphasis on precautionary measures than before. Vulnerability assessments and written emergency response plans (ERPs) are required of all United States water utilities, and utility leaders are required by law to take measures to resolve gaps in security. Many utilities conduct disaster drills to test employee responses and to find weaknesses within their ERPs. Although the increased security measures have led to greater expenses for utilities, the public expects and demands them. Security has also taken a higher priority in other countries.

EMERGENCY RESPONSE PLANS

Most utilities have some form of ERP (formerly called emergency action plans), frequently developed to respond to natural disasters, such as flooding, storms, and civil disorder. These ERPs specify responses and assign responsibilities through all levels of the utility, with a primary objective of protecting the water system, its employees, and its customers. These plans protect life and property, minimize the impact of any event on the utility, and provide for continuation of service. ERPs also identify coordination requirements and points of contact, establish an overall road map to response and recovery, and may identify specific actions to be taken.

Utilities are required to develop ERPs that are keyed to the vulnerabilities identified in their vulnerability assessments. ERPs are developed to ensure a coordinated response to disasters and threats. These plans generally encompass the following elements:

- Specific planning for incidents to which the utility is vulnerable and identification of strategies to assess responses to those incidents.

- A list of resources (internal and external) available for disaster recovery. This list includes personnel, equipment, materials, physical resources, and any predisaster mutual aid agreements or contracts for priority services.

75

- Assignment of clear roles and responsibilities, including who is in charge of operations. The utility manager's primary role in an emergency is that of a resource coordinator and decision maker, usually in coordination with an area-wide emergency operations center.

- Established disaster recovery measures with specific assignments to work teams and managers. This provides a structure for training personnel about their roles and responsibilities and reduces actual response time in an emergency situation.

- Established business continuity measures. Utilities must continue to provide water service, ensure water quality, maintain administrative and financial operations, and respond to customer inquiries. The ERP clarifies how each functional unit of the organization fulfills its responsibilities under emergency conditions.

IMPROVING UTILITY SECURITY

Security issues are not going away. Utility managers are regularly contacted and provided information about new products and services that can help them secure their facilities and improve their selection processes for employees. AWWA Manual M19, *Emergency Planning for Water Utilities*, presents techniques, forms, and principles for developing complete contingency plans. In addition, members have focused on security issues at roundtable discussions, in specialty conferences, and with other utility managers. AWWA and related industry associations, in cooperation with state or provincial and national agencies, will continue to highlight security issues, hold special conferences, and provide essential training for maintaining high alert.

Figure 8-1   Security is more important than ever before to utility managers

In addressing security issues, managers focus on the following important components:

- Utility vulnerability assessments are an excellent road map for managers to improve their business processes while increasing their security. Although technology assists in alerting the utility to problems and may deter damage, attention to the way the utility does business can bring immediate results. How the utility handles cash, stocks vehicles, secures property, allows access to facilities, and selects employees can immediately impact the level of security. Utility managers addressing these issues may experience resistance to change from employees. However, security is critical enough to the utility that immediate changes in business processes may be mandated.

- Expenditure of funds for securing the utility's personnel and property should focus on the three-tiered strategy of Detect, Delay, and Deter.

  — Early (if not immediate) *detection* of unauthorized entry to premises is essential to security, but detection is valueless if it is not correctly interpreted by the individual who receives the alarm information. Utility managers must ensure that they have highly qualified staff members who receive and interpret this information. This may require specific training, specially qualified personnel, the expenditure of funds for outsourced security services, partnerships with other utilities or facility neighbors, or patrol compacts with local public safety organizations.

  — *Delay* mechanisms must be incorporated so that intrusions or potential security breaches are more difficult to achieve. Barriers, reinforced fencing, changes in locks or access controls, safety glass to protect office workers, motion detectors, and lights or sirens are possible delay mechanisms. Some mechanisms may require a public relations effort to explain to facility neighbors that the devices are needed to protect the community water supply. Managers should closely monitor any technology that is added to ensure that false alerts are kept to a minimum.

  — The objective is to *deter* an intruder from reaching the intended target. Vulnerability assessments identify ways to harden facilities so that access to the target is greatly reduced. However, in addition to the security equipment that is added to protect system components, it is essential that employees understand that security breaches can make them targets as well.

- Computer systems have significantly improved business operations, but they bring the possibility of unauthorized access to utility information systems and, potentially, to utility operations. Utility managers, with the assistance of both staff and computer security personnel, should evaluate the various ways in which information and operational data are protected and, as required, allocate funds to strengthen that protection. Customer account information and employee data must be protected, as must operating information, such as maps, site specifications, and SCADA information. Additionally, access to utility information from home-based or laptop computers from a remote location may open an electronic pathway to the utility's information system. Information system security devices are available, as are people who are highly skilled in network protection and security. Employees must be educated about the use of passwords, the dangers of leaving systems in operation while away from work stations, and inadequate system hardware and software protection.

Utilities can spend a lot of money on security. Television cameras, motion sensor devices, extra guards, lighting, alarms, etc., are available from many vendors and may enable the utility to harden its facilities. Improving security at a facility requires that needed resources be identified before an incident occurs. Managers should invest in resources that will meet the utility's needs. While assessing resources, the utility manager should identify safe areas for utility employees' families, training levels of staff, response equipment available, safety and security of facilities, specialized supplies and contractors, mutual aid sources, and local or regional medical facilities that must be served or must serve utility casualties.

But the more important security enhancement may be training and preparation. It is important that utilities establish strong ties to public safety units so the latter are prepared to respond to utility emergencies, as well as communitywide disasters. For example, a chlorine leak at a water treatment plant could require nearly an hour for hazardous materials (hazmat) responders to arrive. If streets are not marked and responders have never been to the remote location, or if responders have never dealt with a large chlorine leak, the situation could be disastrous. Volunteer and professional fire departments need to be oriented to utility facilities and need to know about the properties of chemicals used in the water treatment process. Training exercises held in conjunction with these agencies can expedite their response in time of need.

Managers can increase staff awareness of security by implementing training programs, making personal protective gear available, and conducting practice exercises. Employees can provide managers with great insight into enhancing security because they are the most familiar with facilities and their vulnerabilities, they have the closest working relationship with facility neighbors and (possibly) with public safety personnel, and most take a personal interest in the appearance and reliability of their facilities. Business owners and residents located near utility facilities often notice unusual activities—cars parked adjacent to facilities, people taking pictures, open gates, cut fences, etc. They may notify utility managers if they know who to contact.

Chapter 9 contains guidelines for crisis communications plans. If managers are addressing the public about security or public notification matters, the utility's crisis communications plan must be in effect. All information concerning the matter should flow through the communications professional so that other personnel can devote their full attention to meeting technical and operational objectives. AWWA manuals and handbooks provide more information concerning emergency planning and response alternatives. The key to success is to obtain the material, incorporate it into utility planning and operation processes, and harden facilities and information sources, such as computer networks. All employees must be aware of the significance and criticality of maintaining high levels of security so that the utility meets its objectives during emergency situations, regardless of their cause.

INTEGRATING PLANS AND RESPONSES

It is important that all of the security planning, vulnerability assessments, and prepared response be integrated so the resulting actions fulfill the utility's mission. While each organization must make the choices that best address its needs, a good plan of action is essential and must be easily understood and adhered to by all staff. An initiating event—an intrusion alarm, a physical attack, evidence of unauthorized activity, unusual water quality, threats against the utility, or notification from outside authorities of a potential adverse action—should trigger the following integrated steps:

1. **Assess the initial threat.** Verify whether the threat is valid. Dispatch staff to the site, review alarms, and assess available data.

2. **Activate initial operational response.** Shut down system components to complete inspection and testing procedures.

3. **Determine credibility.** Based on firsthand site inspection, data, and related knowledge, decide whether the threat is credible and what additional action is required.

4. **Evaluate site characteristics.** Determine the need for and extent of staff call-up and notifications to public health agencies, emergency operations centers, public safety agencies, customers, and first responders.

5. **Contamination events.** Initiate sampling processes, expedite testing, request necessary assistance from public health and regulatory agencies, and determine whether public notice is appropriate.

6. **All other events.** Activate the crisis communications plan and ensure accurate and timely dissemination of information.

7. **Protect public health.** At all times, keep your local drinking water regulatory agency informed and involved and err on the side of caution in advising the public of potential boil-water or do-not-use-water notices. It is very easy for managers to get caught up in the events of the moment and not put the public on notice that the safety of their drinking water supply is uncertain. While utilities tend to want certainty before issuing such notices, in contamination cases it may be days before positive analysis is available and contamination confirmed. By that time, significant public health impacts could have occurred, which could have been avoided through prompt and conservative alerts.

RESOURCES AVAILABLE

Security and emergency response matters are an emerging element in water utility management, influenced by legislation, regulations, consumer concerns, quality utility business and operations practices, and common sense. Threat advisories give some indication of potential activities and targets, and utility involvement with local emergency planning and law enforcement planning has expanded the level and quality of security information available to managers. Other sources of information include

American Water Works Association (AWWA)	www.awwa.org
Association of Metropolitan Water Agencies (AMWA)	www.amwa.net
Water Information Sharing and Analysis Center (ISAC)	www.amwa.net/isac
National Infrastructure Protection Center (NIPC)	www.nipc.gov
Association of Metropolitan Sewerage Agencies (AMSA)	www.amsa-cleanwater.org
Sandia National Laboratories	www.sandia.gov
US Environmental Protection Agency (USEPA)	www.epa.gov
Federal Emergency Management Agency (FEMA)	www.fema.gov

This page intentionally blank.

Chapter **9**

Crisis Communications

OVERVIEW

Water utility managers must use all of the principles of communications when a utility faces a crisis that has public impact. Whether the crisis is a major breakdown in the ability to serve customers—a broken transmission line, a water quality problem, loss of a water resource—or a natural or human-caused disaster or terrorist threat, utilities must be prepared for crisis communications.

Every crisis is different and presents its own unique circumstances, obstacles, and response requirements. In general, however, a crisis is an unexpected, major event that has a negative outcome and is more serious than a normal utility emergency. Crises usually disrupt normal business patterns and practices and may threaten the credibility of an organization.

CRISIS STAGES

There are seven stages of crisis communications, as depicted in Figure 9-1. Mature organizations that have anticipated the need for crisis communications begin at step 5 and exercise damage control and proceed to action and recovery. Utilities that have not anticipated such disastrous events often amass and activate significant resources to respond to the disaster while communicating to the public a sense of denial, wishful thinking, possibly anger, and finally a disingenuous attempt to "spin" information. Crisis communications from unprepared utilities frequently begin with public responses that can appear to be excuses, for example, "It's too soon to act," or "It's just an isolated incident." When the public reaction to the perceived excuses is negative, unprepared utility mangers often become hesitant to share more facts. This leads to the perception that the utility is withholding information, further hampering crisis communications.

A utility's crisis communications effort must be a planned action (see Table 9-1). Utilities must be as prepared to communicate during crises as they are prepared to technically respond to those crises. The utility's crisis communications plan should include key messages that will be included in some way in all public announcements, and it should identify target audiences, communications outlets and tools, and protocols. There must be an established implementation plan that includes who will

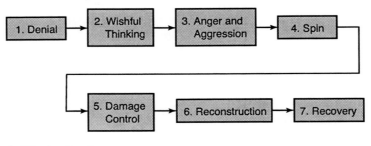

1–4 Dysfunctional
5–7 Mature, highly evolved organizations

Figure 9-1   The seven stages of crisis communications

Table 9-10   Crisis communications checklist

- Assess the situation.
- Gather all pertinent and accurate information as quickly as possible.
- Designate a primary spokesperson and identify other individuals who may have reason to speak to the media.
- Decide what other individuals within the department, elected officials, or other city, county, or state or provincial departments need to be contacted.
- Determine the best way to get information to the media and arrange all logistics.
- Produce all written materials, statements, fact sheets, and updates.
- Determine the best way to reach employees with information on the crisis.

speak for the utility and how media requests will be addressed. In general, utilities are wise to keep all media representatives together, allow them to see and film the same thing, and give them the same access to expert commentary and written materials.

Coordination points must be established so the utility's spokesperson is fully informed about ongoing technical actions and is fully capable of and authorized to deliver essential messages; thus the further need to develop and test a crisis plan before it is needed. The following steps are a proven road map to effective crisis communications:

1. Create a list of potential crises

2. Assemble a crisis response team (see Figure 9-2)

3. Identify spokesperson(s)

4. Develop key messages

5. Determine key audiences

6. Establish a crisis communications center

7. Identify needed technical support

8. Establish communications protocols

9. Build relationships with agencies and organizations that will be involved

10. Develop a media list and become familiar with reporters

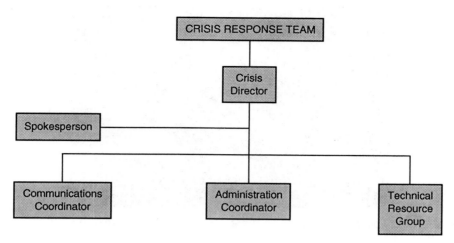

Figure 9-2   Components of a crisis response team

11.   Create fact sheets

12.   Test the crisis communications plan

MEDIA AND THE UTILITY

There are tools to be understood, tested, and used in a crisis situation. There must be a physical location for a communications center. If media representatives are to get the same message and see the same scenarios, utility managers must keep them together and, if appropriate, transport them to the location of the crisis. Technology may allow managers to provide remote video or audio commentary from the crisis location. There also must be a media roster—to call and their telephone, pager, and cell numbers—and a calling tree, including sufficient staff to perform the work. Spokespeople need to have a situation analysis summary for the first presentation and an update template for followup presentations. It is essential that staff make updates in a timely manner. Finally, plan for amenities, i.e., provide appropriate beverages and, if necessary in longer crises, snacks.

When a crisis occurs and a utility representative addresses the media, he or she must remember the important principles of good communications: accuracy, consistency, timeliness, clarity, completeness, and responsiveness. There are no secrets in a crisis; at some point in time, everything comes out and becomes public knowledge. Quality crisis communications address the issues at the time of the event.

The utility must have activated its crisis control center and ensured that all areas of responsibility are fully staffed and functioning. The spokesperson must have a clear understanding of the situation and the concerns that must be addressed and must have access to an on-site source of information. The communications center must determine key audiences—isolated area, community-wide, statewide, national—and the information that will be communicated. It also must establish systems for continual updates and for recording and tracking inquiries. A final update to key audiences is essential—what was done, what remains to be done, what the recovery or rebuilding process entails, and a projected timetable for achieving closure.

Figure 9-3   Always designate a primary spokesperson in crisis communications

The results of these efforts are worth noting:

- There is lower tension because the utility is stating the facts and the actions being taken to respond to the crisis.

- The utility is demonstrating its commitment to open and honest communications.

- The utility is in a position to provide the information and is not reacting to externally generated (and potentially inaccurate) information.

- The utility is securing credibility and building its collaborative reputation.

After a crisis, there is an important element that is frequently set aside—evaluating or measuring how well the crisis communications plan worked. As the utility completes the evaluation process, it is helpful to have a timeline of events in order to see how well strategies or tactics were implemented and whether any gaps occurred between what was planned and the actual event. It also is helpful to identify surprises, including positive or negative events that impacted both the response to and communications about the crisis. The crisis communications plan can be revised to reflect experiences and knowledge gained.

One final task that is part of the utility crisis communications effort is to recognize the participants. A letter of thanks to each media representative for their cooperation is in order. Utility managers may wish to make a token presentation or provide a certificate to reporters recognizing their efforts during the crisis. Internal staff who provided the information that allowed the crisis plan to succeed are to be commended.

Chapter **10**

Human Resource Management

OVERVIEW

Utilities achieve their mission and goals through the work of the men and women who compose their staff. Whether those individuals work on a field crew, in a customer service center, in a treatment facility, or in administrative or management positions, they are an important and costly resource for the utility manager. The key to a utility manager's success lies in the men and women who carry out the organization's mission. A critical managerial responsibility is to oversee the human resources process, including employment, training, performance management, employee relations, and retention of highly skilled and motivated personnel. Whether the utility has its own human resource department or whether it is part of a larger system (municipal or private), there are several key managerial responsibilities.

It is important to remember that a significant portion of any utility's annual budget provides for salaries, benefits, and required contributions for each employee. If that employee leaves the utility, there are direct costs (recruitment and selection, training, overtime, etc.) associated with refilling the position. In an increasingly competitive environment, utilities can experience negative results as a result of inadequate human resource programs. Poor staffing can lead to lower productivity or increasing backlogs, accidents, and injuries with the attendant increased costs resulting from no staff and additional training time.

Human resource issues are also addressed by many regulations and laws, and the utility's management team must be aware of those practices that can result in legal action. These factors include, but are not limited to, fair employment issues, employee relations, safety and security considerations, and attendant regulatory requirements.

The human resource program in any organization encompasses the following elements:

- staffing, including recruitment, interviewing, selection, and all steps necessary to add the new employee to the utility's employment roster.

- training, including general utility matters, specific workplace organization, and on-the-job training

- performance management, including evaluations, additional skills training, and compensation

- employee relations, including counseling and disciplinary processes, employee assistance programs, and termination actions

- policies and procedures that set forth the utility's positions on various human resource issues, such as vacations, benefits, leaves of absence, sickness, equal employment opportunity compliance, employer–employee relationships, safety, and compensation.

STAFFING THE UTILITY

The human resource process begins with a well-planned staffing process. Functional managers should annually review staffing allocations and programs with an eye toward

- probable new program needs and the capacity of current staff to "grow" into those new programs.

- current programs and staffing allocations for programs that are growing, maintaining their size, or being phased out.

- recruitment of the best-qualified candidates for vacancies so as to select individuals who will become productive in their work in a short time and who will be able to work with others in their functional area or within the utility as a whole.

Recruiting the right employee for the right position at the right time is key to a successful utility, and many utilities routinely seek a highly diverse workforce to reflect the community they serve and to comply with fair employment legislation. The staffing process begins with good recruitment and recruitment depends on several components:

- an accurate job description

- competitive compensation and benefits

- a wide variety of sources for potential employees

- a comprehensive understanding of the job and employee market in the area

Many staffing resources are available to managers, including traditional walk-in applicants, job fairs, employment agencies or professional recruiters, nontraditional sources (such as volunteer experience and community service projects), and employee referrals. For promotional positions, an additional resource is current utility employees who meet the qualifications for advancement.

Both external agencies and internal staff rely on an applicable job description. Therefore, the description must reflect the current work requirements; the knowledge, skills, and abilities needed to perform the job; any experience, education, and certification requirements; and physical aspects of the position. To keep job descriptions current, each manager should review all position descriptions within his or her area of responsibility to ensure that they accurately reflect their requirements. There are varying opinions among managers as to whether job descriptions should be highly detailed or very general, but the most important elements are that they are accurate and current.

Utilities have different processes once applications are received for open positions. Initial screening is normally accomplished by human resources professionals, either within the utility or in the larger organization. They rely on the content of the job description to make the initial determination as to whether an applicant's qualifications are suitable to the position. A thorough initial screening and testing (where applicable to the job) process results in qualified applicants being referred for a selection interview, normally conducted by the responsible manager or supervisor or by a selection panel. The objective of the interview is to determine whether the applicant

- is able to perform the work required by the job
- is interested in performing the work
- can work within the organization or work group
- is a good fit for the long-range goals of the organization

There is always some subjective assessment associated with these objectives. However, effective planning for the interview with job-related questions reduces that subjectivity. For examples of interview questions, see Table 10-1.

Interviewing is a manager's primary tool for selecting the right person for the position. It is important to develop and follow an interview plan when analyzing the information submitted by applicants for the position. At the beginning of the interview, managers may wish to give the applicant a copy of the job description so the applicant can review the actual job requirements. In smaller utilities, a single supervisor may conduct the interview. In larger utilities, use of an interview panel is often the chosen interview method. The selection of formal interview panels is crucial in order to create objectivity, equity, and consistency. Interviews should focus on the following points:

- For each key element of the work, ask open-ended questions to get a broad overview of the employee's experience in that area and to put the applicant at ease. The objective is to get the applicant to share information.

Table 10-1   Sample interview questions

Sample interview questions to determine whether an applicant is interested in doing the work.
- What do you know about this type of work?
- What are three things that make this job attractive to you?
- How does this job fit into your long-term career objectives?

Sample interview questions to determine if an applicant is able to do the work.
- Tell me about your experience in (select a key element of the job) and what types of work you did.
- Tell me about the training you received to perform that work and the skills and equipment you used on the job.
- Tell me three things you liked best about that work.

Sample interview questions to determine if an applicant can work within the organization or the work group.
- What are the three most important elements in a good work team?
- What are three things you look for in a work environment?
- What do you like the least about a working environment?

- After the applicant has answered all open-ended questions to his or her satisfaction, ask probing questions to obtain specific information about experience, knowledge, current skill levels, and abilities. The objective is to obtain detailed information as to whether the applicant's skills are what is being sought.

- Finally, ask closed questions (those with yes or no answers) about specific aspects of the job. The objective is to learn what the applicant would do in response to a particular condition in the work situation.

When the manager has sufficient information about the applicant's qualifications, the applicant should be given an opportunity to ask about the job. Before concluding the interview, the interviewer should be certain the applicant has a realistic preview of the work he or she would be doing, including the work conditions, hours, overtime requirements, etc.

After concluding, managers must document the interview. Many managers make notes about an applicant's responses to particular questions to facilitate evaluating them, especially when several people are interviewed over many days.

TRAINING AND DEVELOPMENT

After an applicant is hired, the challenge is to effectively integrate that individual into the workforce and the utility's culture. Training and orientation programs must be organized and correlated with work tasks and performance benchmarks. Also, there must be sufficient time for and documentation of an employee's learning and performance progress. Finally, planning and criteria for ongoing education, certification, skill development, and cross-training programs should be a component of the utility's investment in its human capital.

This element of the human resource management process begins with orientation—both organizationwide and specific to the individual's workplace. Responsibility for orientation is usually divided between the human resource professionals and the employee's supervisor. The human resource professionals give an overview of the organization, important policies and procedures, and compensation and benefits information. In larger organizations that are part of a municipal environment, the utility may provide some type of general orientation about its functions, facilities, and personnel.

Workplace orientation is usually provided by the employee's supervisor. Workplace orientation focuses on specifics, including safety equipment, rules and regulations, supervisor and employee relationships, a tour of the work area and introduction to coworkers, and any printed materials relevant to the workplace.

The job description provides managers with the information needed to develop a training outline for the position, and the training outline provides a way of ensuring the employee knows the work process and of providing needed documentation for the employee's personnel records. It also provides a vehicle by which the employee and supervisor establish performance objectives and standards—clearly stated expectations that form the basis for future decisions about compensation, advancement, advanced training, etc.

PERFORMANCE MANAGEMENT

Utility supervisors and managers have a vested interest in a viable performance management process that includes a review of training results as compared with progress in meeting performance criteria. Usually, performance management focuses

Figure 10-1   Ongoing education and certification programs are an investment in the utility

on annual evaluations, willingness to develop additional skills and abilities, and compensation. It also focuses on the utility's competitive position by reducing the costs associated with turnover.

Ideally, a performance management program achieves these major objectives:

- It helps the utility manager develop current employees' skills and abilities. This includes identifying strengths, improving shortcomings, and charting future development opportunities that support the utility's goals.

- It helps ensure that the utility achieves its organizational goals by including specific, measurable, and interlocking goals that are assigned to specific positions or functions. If goals are not being achieved, the utility can move to a performance improvement strategy.

- It helps the utility manager plan for the utility's future human resource requirements through an ongoing inventory of employee skills and abilities. This enables the manager to know who is capable of doing what and how capable they are, employee interest in and potential for self-development, and employee readiness for advancement.

Performance management programs must include an evaluation process, usually a formal written dialogue between the employee and his or her immediate supervisor. Most utilities have, at a minimum, an annual performance review for all employees. The goal of an appraisal process is to give feedback on performance and work-related behaviors and to plan development steps to increase the employee's value to the utility. It is important that each performance appraisal interview be planned by the supervisor(s) involved so that it addresses the work performance, improvement needs, development objectives, and areas of interest or concern to the employee. Sufficient time should be allocated for the actual interview, and it should be designed with consideration of the individual employee.

While appraisal systems help managers determine whether additional compensation is deserved, their primary purpose is to assess work performance and to develop new and improved job skills. Compensation decisions and performance management system appraisals should be separated at the earliest point possible in an employee's career with the utility. Utility managers must advocate for adequate compensation in larger organizations so that utility employees' responsibilities for ensuring public health and safety and for maintaining environmental quality are sufficiently valued.

Successful performance management systems are based on clearly defined objectives and procedures, conducted by trained supervisors and managers and applied consistently throughout the utility. The organization's senior management is responsible for ensuring that all persons responsible for appraisals are trained and capable of conducting such reviews. Senior management must also review the performance management system and make necessary adjustments to meet the utility's goals.

Some organizations have adopted an appraisal process in which an employee rates his or her performance and also receives performance feedback from a peer, a superior, and (if appropriate) a subordinate. This type of feedback provides significant insight into how the employee's work performance is viewed by coworkers.

EMPLOYEE RELATIONS

Managers are responsible for establishing a workplace environment that respects diversity, promotes equal employment opportunity, encourages open and honest communications, rewards excellence in performance, and values each employee's knowledge, skills, abilities, and contributions to the organization's goals and objectives. Managers set the tone for high-quality employee relations through ethical behavior and commitment to developing their personal knowledge and skills. Employee relations is that area where utility management is continually assessed by employees as to whether there is a difference between management's stated policies and procedures and management's real actions.

Institutional ethics is an area of great concern to constituents, managers, and employees. Utilities must take all steps necessary to develop and enforce conflict of interest and ethics policies and procedures. Employees who see managers with impeccable ethical standards emulate that behavior.

Open and honest communications are the foundation of quality employee relations. Managers must be aware of the needs and expectations of employees and their work activities and must foster an environment that encourages and celebrates communication, innovation, and excellence. Regular meetings with employees are an easy way to communicate organizational goals and objectives and to gain feedback. Employees are an exceptional source of knowledge regarding concerns or areas needing action in the greater community, but they need to know that their information is valuable to the utility and its future. Regular meetings with the managerial and supervisory staff ensure that the utility's goals and objectives are kept in the forefront and that progress toward achieving those goals and objectives— or redefining them, if necessary—is shared and then disseminated through the organization.

Many utilities have employees who are members of one or more labor unions. A hallmark of good employer–union relationships is the ability to establish and maintain communications on work processes and potential changes that will impact the utility and its employees. In general, there are many areas of commonality that form the basis for quality employer–union relationships—safety; fairness and equity in supervision, work assignments, and rules; and problem-solving regarding equipment, staffing, and training are among the most familiar. Managers who succeed in building good relationships with unions have developed mutual trust with union representatives and employees and have taken extra steps to avoid developing adversarial relationships. This does not mean that there is agreement on all issues; nor does it mean that there are not issues where each side takes a particular position that is adversarial. It does mean that there is room for discussion, negotiation, and collaboration in such matters rather than a swift descent into acrimony.

Much is said about the need for utility managers to maintain open-door policies that encourage easy communication with employees. Managers must take care to ensure that their open-door policy does not become a conduit for organizational dysfunction. In many performance management approaches, managers become a point of appeals for adverse actions, and it is important that they be viewed as objective on such occasions. The perception of favoritism or unlimited access works against positive employee relations. Many concepts of excellent management appear each year, and managers can examine and possibly add them to their toolbox. A tried and true principle continues to be used—periodically being physically present at worksites throughout the organization. It affords an opportunity for accessibility to employees and their concerns without seemingly giving greater credence to one work unit or to an individual.

There is also a need to provide employees and work groups with tangible evidence that they and their work are truly important to the utility. Whether this is regular recognition of positive performance efforts, a utilitywide celebration of a single, notable achievement, a customer commendation or recognition of long-term employment, it is important that managers demonstrate their appreciation of employees' efforts. Such events are not expensive or difficult to schedule and conduct. But they must carry a message to employees that the organization and its managers are aware and supportive of employee contributions.

DISCIPLINARY ACTIONS

There are instances where managers must address performance or behavioral issues that do not comply with the utility's policies, standards, or expectations. Unless a particular incident is so egregious as to require immediate punitive actions, most utilities provide for a structured progressive discipline approach in handling such issues. This concept is based on the principle that managers and supervisors are closer to the employee and the problem and are the ones best able to assist in resolving the issue. A key element in disciplinary actions is for managers and supervisors to be alert to problems and intervene as early as possible. The process works best when managers and supervisors are trained in the counseling skills necessary to make interventions. It is important that managers and supervisors are aware of community resources that can assist them and the employee. Upon identification of a problem, the affected manager or supervisor is responsible for calling the responsible employee's attention to the problem, providing counseling about the matter, and working out an improvement plan and review date. Discipline procedures differ, but normally include (in order of severity)

- identification of performance or behavioral issues and a counseling effort to correct them

- an official oral reprimand with supporting documentation

- a written reprimand describing the problem, the efforts to correct it, and establishing a time period for resolving the issue and the attendant consequences if it remains unresolved

- a suspension for a designated period of time, usually without pay and, often, with a reduction in benefits

- termination of employment

Such disciplinary actions are taken after investigating and documenting the performance or behavioral problem and then conducting an appeals process through succeeding levels of management to offer an opportunity for review of the decision processes. Managers must ensure that this appeals process is not stifled and that an unbiased review is accomplished at each level.

There are also instances where the utility manager must take action to assist an employee who is unable to meet performance objectives because of health or personal reasons. Larger utilities may have their own employee assistance programs to help employees deal with personal issues, while smaller utilities form partnerships with local service providers and refer employees to those agencies and their professional staffs. Utility managers must have the knowledge and skills to recognize that work performance or behaviors have changed. In addition, managers must have the ability to engage the employee, build trust, and assist the employee in gaining access to professional assistance. In general, managers take the role of initial intervention and referral; the professional partners take on responsibility for dealing with specific actions to remedy the situation.

Finally, all managers must engage in lifelong learning to acquire the knowledge, skills and proficiencies that enable them to be strong leaders for their utilities. Rapid technological advances, changing economic conditions, resource limitations, the global business environment, and many other trends demand that managers anticipate future challenges, then acquire and apply knowledge and skills to address them. This personal demonstration of willingness to grow sets the example for others that management staff must expand their own education level and technical competency to be sufficient and remain competitive.

POLICIES AND PROCEDURES

To ensure consistency across functional areas and management levels, utilities develop and comply with human resource policies and procedures. In cases where the utility is part of a larger municipal organization, it is governed by the policies and procedures set forth for all employees. At the same time, the utility may need to develop and implement policies and procedures that are specific to its operations, such as those relating to operator training and certification, customer contacts, and utility-specific work conditions.

It is critical that policies and procedures be dynamic documents and remain dynamic in changing employment environments. Far too often, policies and procedures are codified and filed away in case they are needed in some future circumstance. The result is an out-of-date document that does not reflect the utility's new and different situation.

Managers must have all policies and procedures regularly reviewed by the organization's senior management staff to ensure that they reflect current laws, utility customs, and performance requirements. Managers must also ensure that all levels of supervision are familiar with the content of the policies and established procedures and that they know and apply their roles and responsibilities as part of the management structure. The recommended regular review process allows the utility to eliminate policy statements that are dated; revise those where changing conditions or laws require modification; and develop new statements where necessary. Incorporating the review process with the training process brings a broader evaluation and a better understanding of both the policies and managerial responsibilities for their implementation.

SUCCESSION PLANNING
AND PROFESSIONAL DEVELOPMENT_____

An essential element in successful utility management is ensuring that employees are prepared to move into a higher position within the organization. That requires an investment in human capital—everything from obtaining high school equivalency or college degrees to attending technical training courses, industry seminars and conferences, managerial development courses, cross-training, and in-house skills programs. Most performance management programs require the employee and his or her supervisor to identify a personal development component that becomes part of the individual's objectives for the coming evaluation period.

Utilities that budget for and expect achievement in the areas of professional development reap three benefits:

- There is a larger body of multiskilled employees that enables the utility to respond more quickly to emergency situations or to reallocate staff when needed.

- The utility and employee can document increased knowledge, skills, and abilities so that when advancement opportunities arise, there will be qualified staff capable of moving ahead in terms of responsibilities and rewards.

- If there is a need to restructure functions or reallocate staff in light of competitive issues, the utility is better prepared to take such actions based on bona fide criteria and factual information.

People skills are an essential element of good management but are not the easiest part of the manager's job. While the demand for administrative activities occupies much of managers' available time, managers find active involvement with staff throughout all organization levels to be both informative and rewarding.

This page intentionally blank.

Chapter **11**

Information Systems and Services

OVERVIEW

This chapter focuses on information technology (IT) and information systems (IS), which have an ever-increasing role in any utility manager's toolkit. This chapter gives managers a broad understanding of the complex issues that accompany the implementation or extension of IT within their organization.

Utility managers can have an important tool in IT and IS, if they effectively use information systems and services. Expanded business systems have found related applications in all segments of utility operations. Billing, customer service, and accounting applications have been joined by supervisory control and data acquisition (SCADA) applications to remotely operate pumps, tanks, and other appurtenances. Automated meter reading has been supplanted by computer-assisted, drive-by reading systems. Field work orders, mapping, and inventory acquisition can now be accomplished by operations work teams while at the worksite. All of that information is available for analysis by managers with the objective of improving utility effectiveness and efficiency. However, managers are in the unenviable position of having the benefits of and information from automation without the ability, often, to fully utilize the power of those benefits and that information.

Utility IS are complex mixtures of machines, instructions, and people. And like many technical fields, the IS arena is filled with jargon. The good news for managers is that the jargon refers to machines, instructions, and people. Managers can best understand the role of systems in the utility organization by considering technology as a series of building blocks that form the solution to an information problem or situation.

Oversight and management of a utility's IT should be high enough in the organization to reflect its strategic importance. Usually, the mission-critical nature of IT management is important enough for the utility's IT manager to be part of the organization's senior management staff. Skills acquisition is an ongoing process for both IT professionals and user groups and is largely the result of continuing improvements in both hardware and software. Managers should be aware that IT professionals are in high demand both within the larger municipal sector and in the general business environment. High turnover rates are a continuing challenge and retention strategies must be a part of the manager's toolkit.

KEY ELEMENTS OF IT

The building blocks that make up the modern information system are

- the computer and its related components—the *hardware*

- instructions used by the computer to accomplish its assigned task(s)—the *programs* or *software*

- the business environment, often referred to as the *problem space*

- the technology environment, referred to as the *solution space*

Role of the Computer

The role of the computer is deceptively complex until its basic functions are understood. In any information system, the computer performs three functions: it performs computations, collects and stores data, and organizes and stores information. While computers appear complex, they actually are simple devices—they perform calculations, search processes, and provide reporting functions, but they can only perform one instruction at a time. What makes them amazing is that they perform these functions at high speeds—often billions of instructions per second. Further, computers are capable of capturing the results of the activities performed by an organization through *data entry* functions. Captured data is stored in *files* and *databases* until needed by the system user. Computers then use the data stored in files to respond to requests for information relayed by an application program that also provides instructions on how to organize stored data into a context the user understands. The organized data requested by the user is relayed in a report or displayed on a monitor so it can be used to solve the identified business problems.

Computers are the most visible component, and IS often are identified with the hardware, rather than the computer being identified as an element of IS. So much emphasis has been placed on computer hardware and not enough on other elements of IS that many system deployments end in disillusionment and, in some cases, outright failure.

Role of Software

Software is the instruction set (program) that enables the computer to perform its operations. There are three basic categories of computer software used in IS construction:

- **Operating systems.** This software is loaded when the computer is turned on and provides the computer with an instruction set for functioning. Managers are familiar with the common operating systems in use, including DOS, Windows®, Unix, MVS, and Linux.

- **Application software.** These are computer programs that perform specific tasks or solve specific problems. These are the programs that are used most frequently, with common examples being billing programs, word processing programs, spreadsheets, and presentation programs.

- **Database management software.** This is specialized file-management software that captures and stores data and organizes and reports information to support the user's decision-making requirements. Not all systems use database management software. Instead, they rely on the operating system's file management capability to store and manage data files. However, as data storage requirements and reporting needs become more complex, a database management system becomes necessary.

Impact of Current and Future Business Environment

The pressures of increased globalization, stresses of economic change, rapid expansion of urbanization, evolution of marketplace demographics, increased levels of competition from unexpected sources, and new expectations generated from both successes and failures all pressure organizations to change everything about the way they are organized, do business, and make decisions.

The business environment is the *problem space* for information systems. This is where needs for decision support and decision-making information are defined with regard to the interaction and pressures of the internal and external forces at work in the organization. Changing social, economic, political, and technological forces affecting the utility will shape its future business model and environment. A utility's future success will likely be determined by how well its IS setup allows it to anticipate and react to change and how well the organization's decision makers are able to use technology to solve business problems. Technology issues and IS will play leading roles as utility managers direct the organization's resources to accomplish its mission.

The Technology Environment—Current and Future

In the early days of computer technology, the common approach was to acquire computer resources, then seek applications that could be automated. As the benefits of IS became more widely accepted, managers began to develop the solution first, then acquire the technology appropriate to the design. Thus, the technological environment has become the *solution space* for many of the organization's problems. As a maturing component of organizational efficiency, IS designs now provide tools for rapid decision making and solution deployment, spawning a marketplace in which computer technology becomes obsolete very quickly. From a managerial planning and budgeting perspective, many analysts suggest that cost recovery of technology investment should be done on a 30–36-month schedule. This becomes even more feasible when considering the relative cost stability that results from technological advancement. The need for faster computing and the technology response are present in every electronic medium and will continue to drive both business and personal IS requirements.

IT STRATEGIES

Utility managers must demand a good requirements definition of the business environment and a carefully documented and communicated plan for each use and deployment of technology. They must provide as much time as needed in the planning process for implementing IT and IS in order to address human concerns and reticence. IT strategic plans must provide utility decision makers the ability to take advantage of the rapid advances of technology, as well as the capacity to exploit fast-paced advancements in IS capacity. They also must provide the internal framework to communicate, integrate, and merge methodologies and competing goals into a coherent guide to success.

Developing and Implementing a Strategic IT Plan

Once again, a process enables the utility manager to plan, organize, direct, and control the development and implementation of a strategic IT plan.

1. **Establish clear goals for the IT planning process.** The managerial team should identify stakeholder expectations for IT implementation and use those expectations and business needs to establish priorities for use or expanded use of IT. An ongoing objective is to create broad cross-organizational understanding of and wider involvement in IT issues and planning processes. Goals should also include identification of projects that are needed for organizational success, resources needed to implement IT solutions, and a plan to develop a workforce that is IT capable.

2. **Establish criteria for IT strategic planning.** Such criteria must consider the utility's business vision and result in output that has meaning to the organization. It also should provide a basis to resolve conflicts, prioritize information needs, and align competing projects. Ideally, it requires a rapid development and revision process and results in a way to develop coordinated operational plans.

3. **Agree on contents of the plan.** The plan should be as concise and clear as possible. An example of its content appears in Figure 11-1.

4. **Ensure a relationship of the IT plan to the utility's business strategy.** Remember that the IT plan is a part of the business strategy because it defines how information technology is used as a decision support for all functional units. It should define resource allocations that support achieving the utility's strategic vision and accomplishing its mission. At the same time, it must establish how it supports operational objectives such as cost reduction, less exposure to risk, improved organization performance.

5. **Give serious attention to ancillary issues that will impact the organization.** These include, but are not limited to

   a. assuring that the corporate and organizational unit visions and missions are understood and accepted

   b. determining the utility's willingness to accept technological risk—is the utility a "bleeding" edge, a leading edge, or a follower?

   c. allocating funds, staff, and administrative support for the project

   d. providing adequate professional project management support and widespread understanding of the project methodology

   e. establishing a realistic timeline for the project

   f. ensuring that technology is available and staff understands the use and deployment of technology

I.   Management Summary

II. Introduction
     A. The intent of the strategic plan: audience and purpose
     B. Level of coverage and relationship to business unit and other
        plans

III. Enterprise Mission and Strategy (Business Thrust)

IV. IS Mission and IT Vision (Where to)
     A. Primary role of IT for the enterprise
     B. Visions of IT contributions in business terms

V.   IS Strategy (How)
     A. Goals, needs, and opportunities for IT
     B. Current environment
        1. Systems, networks, and applications (use portfolio analysis)
        2. Stance versus the competition
        3. Technology products on market
     C. Resources available (budget, staff, and contractors)
     D. Path and means to achieve objectives; statements of direction
     E. Criteria for ranking projects

VI. Strategic Objectives (What)
     A. Share of resources by portfolio class
     B. Utility class objectives (mandatory, cost reduction)
     C. Enhancement class objectives (improving corporate performance)
     D. Frontier class objectives (high-risk, high-potential trials)
     E. Infrastructure objectives

VII. Implementation Program (When and How Much)
     A. Allocation of funds, staff, and the timeline
     B. Responsibility, architecture, and project management
     C. Plan for review process and final evaluation

Figure 11-1   Outline for an IT strategic plan

IT AND IS INFRASTRUCTURE

Information technology architecture (ITA) provides the basis for managing and controlling the quality of IS, with a basic idea of promoting managed change and migration and lowering the total cost of IT ownership. It describes the relationships between the work of the organization, the information used, and the technology employed to perform the work and process information. It also assists managers in limiting duplicate systems or those that perform overlapping functions and can provide the technical vision for guiding cost control and performance improvement.

The technological infrastructure is the description of the functional components, features, capabilities, and interconnections of the IT system and includes, as its major components, hardware, software, network and telecommunications links, and relational database management systems (RDBMSs).

- **Hardware.** Think boxes and attached devices—the computer with its hard disk drive and CD or "floppy" drives; monitor, keyboard, mouse, printer, modem or other network connection, scanner, etc. This is the part of IT where the machine and the human make contact.

- **Software.** This includes the operating systems (loaded at computer startup) that are essential for all computer operations and resource management and that control the human access to the monitor and other computer peripherals.

  - *Applications programs.* These are programs that perform a specific function for the user. They may be developed within the utility or purchased from vendors who specialize in a particular application. Major utility applications include CIS, work management and maintenance management, GIS, human resources, financial information, document management systems, SCADA, fleet management, regulatory reporting, and specialized technical uses.
  - *Security programs.* These are special applications that provide protection for systems and networks from hackers, unauthorized users, and attacks on physical and software systems.

- **Network and telecommunications.** Networks are a series of points (single computers or other networks) interconnected by communication paths. They can carry voice, data, or both types of transmissions. They usually are classified as local area networks (LANs) (serving one location) or wide area networks (WANs) (serving multiple locations through a network of networks). They are able to communicate with each other via a transmission technology called a protocol. They may use utility telecommunications resources, such as the telephone system, dedicated data service lines, or cable television networks.

- **Servers, server software, and connections.** Servers are computers that control access for the network, monitor storage, and distribute applications among users. Servers require a server operating system that controls multiple resources. Connection options include modems, wireless, high-speed ethernet and broadband. Utilities should choose the options that best meet their needs and capabilities to service the organization's IT strategic plan.

- **Relational database management systems.** These are collections of interrelated data stored together and organized in a manner that benefits multiple applications. The most commonly used are RDBMSs, which control storage and retrieval of information, keeping data accessible but organized as intended. With RDBMSs, multiple users can access the same data at the same time. An emerging database system is the enterprise data model (EDM), a high-level, consistent definition of all data developed and used across the entire scope of the business that focuses on how each data element relates to all other data in the enterprise. Both approaches require continuing administration to ensure they are maintained. It is worth noting that the EDM defines strategic information need, but that modeling the enterprise is a relatively slow and labor-intensive process.

WEB TECHNOLOGY

Web technology is increasingly used by utilities to provide customers and the public with information about services, programs, and important milestones in operations. Growth in Internet usage has been phenomenal, with usage becoming more widespread in the population as technology itself becomes more pervasive in the workplace. The World Wide Web is a network of computer resources and users using the HTTP to access available information. With the growing demand for 24-7 information availability, utilities have responded with customer services and tools that meet those expectations. That same demand for information availability is present in the workplace, and workgroup collaboration is more easily enabled using Web resources, such as instant messaging and intranet.

When implementing Web technology, utility managers need to keep four factors in mind:

1. Someone must be in charge. This individual must monitor, change, and update Web sites on a regular basis.

2. The Web site must be secure. If the Web site is used for payments, opening and closing services, accessing customer information, and other customer service functions, security is paramount; it must prevent access to personal information by unauthorized parties.

3. The Web site must enhance the utility's image. Information should be appropriate, accurate, clear, up-to-date, and useful. A Web site that was last updated in 2000 does nothing for the image of the utility.

4. Web use within the utility must be monitored, and acceptable use standards regarding the personal use of e-mail and the Internet must be established.

MANAGING DATA

Data management is the key ingredient to successful use of IT. In general, the model for data management is *capture, store, retrieve, organize,* and *report.* This allows data to be reorganized much more easily and affords multiple users to see many different views of the same data. Maintaining data integrity is a function of three elements—data quality, data ownership, and data integration.

Data quality is often equated with accuracy but is better described as data that fit the intended purpose for the decision-making needs of the utility. In order to be used, data must be accurate, timely, complete, consistent, objective, and accessible. It must be relevant, available in the appropriate quantity, concisely presented and easily understood, and have a consistency of interpretation. Data management software must assure that the right data get to the right place, in the right context and format, and at the right time. The cost must be low enough that decision makers can accomplish the mission and objectives of the utility. Emphasis on data quality ensures that

- data collected reflect what was intended in data definition

- data in any table of the database are properly integrated into the overall enterprise data model

- data are available for user needs within a reasonable amount of time

- data are available to serve the multiple needs of users and allow each to make intelligent use of the data

- data are able to provide value to the decisions made within the organization

Data ownership is generally assumed to be an organization asset, rather than the property of the functional unit that produces the data. This approach reduces or eliminates the duplication of data stores, decisions made from different data sources that may contain variant information, and bad decisions resulting from divergent data. A critical element in defining data ownership is assigning responsibility for ensuring that data assets are available to decision makers so they can do their job effectively, using the best information available. Thus a requirement within IT is to assure that an individual or organization unit is responsible for making the data definition, determining access rights, defining data quality and updating data, and establishing information processes, packaging, and delivery.

Data integration is the process of reorganizing data from disparate systems and incorporating it—with a minimum of redundancy—into a database that serves enterprise applications. The integration of data is supposed to create new operating efficiencies and improved business processes.

SUPPORTING IT

Effective use of IT requires a quality support structure within the organization. That support may be provided by utility employees or by third-party vendors. At the very least, IT requires a help desk—a designated person or site users can contact for assistance and to get answers to questions. Many small organizations use knowledge-able staff members as their help desk or problem-solvers, but a true help desk usually includes a group of experts who use special software to analyze and track the status of problems, take requests for technical assistance, or dispatch appropriate personnel for major problems or repairs. Help desks are valuable for many reasons. They

- minimize user productivity interruptions

- provide a unified approach to solving technology problems

- provide a central point of contact for routine questions or major repairs

- track problem areas that may require software changes, user training, new procedures, etc

- deliver customer support in a unified manner

Some utilities outsource IT support services rather than provide staff to address problems. The common motive for outsourcing is usually economic—utilities believe they can have the same services at a lower cost from an external supplier. Other reasons include access to cutting-edge technology, skills, and knowledge that enable organizational change while solving IT problems; ability to provide short-term resources without adding long-term costs; and ability to focus resources on core strategic IT goals. At the same time, outsourcing poses potential problems, the most critical of which is that the vendor's business goals are different from the utility's business goals. If the utility becomes dependent on the vendor, it can be extremely expensive to move to another model. Also, there is the probability of a communication gap in building mission-critical IT solutions.

If the utility is seriously considering outsourcing IT support, some areas are more suitable than others. They include, but are not limited to, network and network maintenance, telecommunications, Web site development and maintenance, help desk, intranet, application development, security services, and call center support.

Utility managers may choose to establish a service level agreement to provide support for the utility's IT program. It is a contract that defines the technical support or business services to be provided by a vendor and it spells out service expectations, service performance measures, and the consequences for failure to meet service levels. A good service level agreement should

- provide a clear statement of purpose
- describe in specific terms what will be provided
- define the delivery mechanism
- define the metrics used to assess service levels
- define penalties for lack of performance
- specify the change management process
- define installation timetables
- describe termination conditions and legal issues

STANDARDS

Standards are benchmarks that define the required function level for hardware, software, network, and other IT products and services. They are important because they form the basis against which hardware, software, systems, and data quality are controlled and measured. They also specify the parameters of future acquisition decisions. Standards range the continuum from proprietary vendor systems, such as the ones embodied in various software releases, to the standards of the International Organization for Standardization and the American National Standards Institute standards organizations. Between the two extremes are standards set by specific product consortia.

Utilities may write their own standards or adopt those of standards organizations (e.g., ANSI, AWWA). Utility-specific standards are usually developed in-house and reflect the general industry standards. Usually, standards are established for hardware and software, Internet/intranet, and networks. Regardless of the standards developed, they must be

- realistic and specific
- measurable and verifiable
- consistent
- clear and understandable
- dynamic

IMPLEMENTING NEW SYSTEMS

Implementation of new systems includes both those developed internally and the various processes involved in converting to a commercial system. While the tasks in the process may differ, the methodology is essentially the same. Top management must support the process and the process itself must have a clear vision of objectives and a carefully developed methodology for introduction and implementation. Also required is a realistic timeline to complete implementation tasks and ensure system quality. Tangible products or deliverables should be defined for each phase of the project. There are seven distinct phases in the process of implementing new systems:

1. **Initiation.** The utility requests assistance in solving an IT problem. The scope of the problem is studied and a working problem definition is developed.

2. **Requirements definition.** A description of the requirements and the user's expectations of the resulting system are developed.

3. **Analysis.** A detailed definition of the system is created that develops functions specifications, a project plan, and (if required) a request for proposal (RFP) to be sent to vendors who will supply the desired system or parts thereof.

4. **Design.** A blueprint of how the system will be put together is developed.

5. **Development.** The system defined in the design documents is created. It includes the design and coding of programs that will input required data, store and process that data, and deliver the information in the specified context. This phase also includes testing all elements of the system to be sure they function as requested and readying the system for implementation.

6. **Implementation.** This step brings the project to life for the utility, completes the final integration testing, and takes the new system online. At successful completion of this phase, the user accepts the system.

7. **Maintenance.** This phase monitors the implemented system to ensure that it performs as specified. Modifications to correct deficiencies or incorporate changes for overlooked functional requirements may be necessary. While maintenance is not always included as part of the process, it is an essential phase in ensuring that a quality IS is received.

Build or Buy

All system development projects reach a point where utility managers must decide if the utility IT group should build its IS new or instead purchase the IS complete or near-complete. The answer becomes a strategic decision with long-term impact on the utility—encompassing financial, operational, technology, and customer support elements. The best answer to the question is found after the utility

- analyzes its vision, mission, and strategic goals *vis a vis* IT

- determines the adaptability of its business rules and process to fit a system that is not custom-tailored

- assesses the availability of internal technical resources to construct the system on schedule and within budget

- assesses the availability of its organization to provide long-term support in-house

- assesses available systems, products, vendors, support organizations, and developers in the marketplace

- assesses the availability of commercial off-the-shelf systems that have the features and functionality to satisfy its user requirements

- evaluates the direct costs of developing, operating, and supporting the system, as well as the opportunity costs of diverting resources from its core businesses

Whether building or buying a system, the utility must manage its implementation, thus project management expertise must be available either from staff or from consultants.

Lease or Purchase

While the build or buy decision involves the strategic decision framework, the lease versus purchase decision is primarily a financial decision and relates essentially to hardware. The decision is based on a cost–benefit analysis, assuming a lease contract and an installment purchase contract of equal or similar terms. Utility managers should consult legal advisors involved in the process to ensure that all bidding, procurement, and tax implications for the lease contract are observed. Critical decisions include the utility's choice of vendor, negotiating and managing the lease contract, and safeguarding leased assets. The following are some questions that should be answered:

- Does the utility want to own the assets at the end of the term?
- Is equipment replaced on a regular schedule?
- Can all of the equipment be located at the end of the lease?
- What is the organizational disposal policy?
- Does the organization have an asset management program?
- Has the organization ever leased assets before?
- How long are assets to be leased typically used?
- How rapidly is the technology under consideration changing?
- Is there a need for quick adoption of new technologies?

Selecting a Vendor

Once an outsourcing or purchase decision is made, the vendor selection process begins. The formal process should always begin with an RFP that sets out in clear, specific terms exactly what product or service is being solicited and exactly how the successful vendor is evaluated and selected. General criteria include

- the vendor's experience (including experience with projects similar to the one in the RFP)
- product features and functionality
- technical architecture and specifications
- speed of deployment, installation, and conversion
- market research regarding the product under proposal
- planned responses to emerging technologies
- the vendor's market position
- short- and long-term cost of ownership of the proposed product
- proven ability to deliver as promised
- industry-specific knowledge and experience
- vendor resources (such as employee strength, experience, and training)

- maintenance and support capabilities
- the vendor's ability to work with other systems providers (if the utility uses a multivendor platform)

Managers use a checklist, questionnaires, comparison matrix, or other type of scoring for evaluating all vendors. This same approach is used in evaluating software and other products.

Adapting Business Processes

When using software systems supplied by third-party vendors or other outsourcing firms, the utility needs to adapt its business processes and rules to fit the software. Vendors may make some accommodation for business model exceptions, but the major responsibility for adaptation falls to the utility. Adaptation of business processes should provide the utility a competitive advantage from improving and integrating business processes. Managers should include adaptation of current processes in developing the user requirements definition, because this is where acceptance of the new solution begins. Not all process adaptation is negative.

Many applications are employed with the intent of reengineering core business operations. These applications are enterprise models and require a shift in managerial thinking from business units to an enterprise model. In short, to build an enterprise model utility, managers should concentrate on integrating departmental needs throughout the organization rather than identifying and focusing on the narrower specific functions.

Data Migration

Data migration, also called conversion, translates information from the current applications or database into a new format required by the new application or database. Its purpose is to protect data related to the core business and make that available through the new application. Migration is a complex process and often is a source of great stress on the organization. Conversion planning should begin as soon as possible after implementation decisions are made. The conversion planning process should assess the number of systems involved, volume of data, data mapping requirements, number and format of source files or tables, data transformation staging and data-cleansing requirements, and the availability of data transformation tools. Data conversion is never easy, and it is never accomplished quickly.

Implementation

Implementation occurs when the project is almost complete and all of the systems elements are ready for assembly into the IS. However, tasks in the implementation phase are often the most overlooked part of the project. Some are rushed or omitted altogether, and many system developers consider tasks such as integration testing or user training as redundant or unnecessary. This generally results in customer dissatisfaction. Managers must be aware of the following major tasks in implementation and insist that they are specifically identified in any developer contract:

- installing hardware, network, and software
- completing data migration and loading databases or files
- running an integration test suite to assure that all elements work together
- completing user training

- running acceptance test suites
- correcting problems found in testing activities
- re-running integration and acceptance tests
- user and managerial sign-off
- turning over operations

Ongoing Support

For purchased software, ongoing support is generally provided under an annual maintenance agreement that provides help desk and error correction support. It also usually provides for software updates as the vendor improves and corrects the program's functions. For the internally developed system, the utility's IT staff provides these services. Modifications may be required because of errors that were not caught in the testing phase, changes in business rules or operations or in government regulations, system shortcomings, and changes in customer requirements.

EMERGING ISSUES

Several emerging issues will have a direct impact on IT and cause the strategic decisions that utility managers make to be even more critical. They introduce a level of risk and, in some cases, uncertainty into the decision environment. They are not controllable by the average business entity, but they will constrain the decision process.

- **Rate of change.** The rate of technology change is exponential and impacts the decisions that must be made. They include, among others, replacement policies and asset management programs, training requirements to update employee skills, increased investment from shorter life cycles of technology products, and shorter management change cycles.

- **Technical obsolescence.** Obsolescence affects all technology resources—people, hardware, software, communications, data, and business processes. It cannot be controlled or eliminated, but it can be incorporated into the utility's decision model as a risk factor and quantified as any other risk factor.

- **Changing business processes.** Core business operations of the utility are continuously evaluated and frequently reengineered, probably shifting more toward an enterprise framework of planning and execution as resources become scarcer and more expensive.

- **Changing education level of the user base.** A utility's future employees and customers will be more technology literate and will have a higher expectation for the availability of technology solutions. Education has provided computer access to students, while creating an uphill battle for businesses and institutions that must employ and serve adults.

- **Measuring the effectiveness of the IT investment.** While IT investment effectiveness has traditionally been measured by financial metrics, future measures will include the impact of opportunity costs as well as direct costs—value to the utility. Some potential criteria include customer need, strategic fit, revenue potential, business and technical risks, the level of required investment, and the amount of innovation and learning generated.

Technology should not be deployed without an IT strategic plan. That plan must interface with the utility's strategic and capital plans. Utility managers must realize that technological progress is occurring exponentially rather than following a linear model. This rapid change means that managers must anticipate and deal with built-in obsolescence. A generally accepted rule of thumb is that investments in technology should be returned within three years. While decisions seem complex, managers must bear in mind that they support the operational plans, requirements, and decisions of the utility.

Chapter **12**

Legal Issues

OVERVIEW

A utility's attorney may be a manager's best asset. The broad range of policy requirements, contracts, legislative and regulatory issues, personnel actions, customer activism—not to mention possible litigation—leads to the conclusion that advice from counsel helps the utility manager. If the counsel's advice is identified at the earliest stages of response to an event or project planning, it can save the utility significant costs in time, stress, and dollars.

The role of an attorney is not to make management decisions. It is to know the utility, its mission, and legal environment for operating and to give counsel when called on. Managers are responsible for seeking advice when it appears that legal support may be needed. In addition to the familiar function of the utility's attorney as an advocate in litigation, enforcement proceedings, or alternative dispute resolution, counsel may perform many other tasks that are directly related to the general provision of utility services. These include, but are not limited to

- forming public utility agencies
- contract negotiation and drafting
- regulatory guidance
- pursuing permit applications
- rate consultation
- legislative services

SELECTING THE APPROPRIATE COUNSEL

The manager's primary objective in identifying legal counsel is to obtain the most qualified legal help at a reasonable cost and to take immediate steps to integrate legal counsel as an effective and accountable team member. Utility managers should select an attorney, possibly contract for outside legal services, and create a good working relationship with counsel through effective communication.

Attorney Selection

The first step in identifying appropriate legal counsel is to clearly identify the utility's need for legal services. Managers should consider whether there are particular legal issues associated with a matter or if a project involves unusual legal questions, such as those related to endangered species or designated wetlands. Managers must also consider what legal experience the utility needs (i.e., litigation, administrative law, contracting) and whether past experience with the utility is of benefit. Other factors weighed include whether there are geographic or regional issues, whether the utility requires experience working with a particular regulatory agency, whether there is likely to be a public protest leading to a contested case or litigation, and whether political visibility is desirable. The utility may be best served by in-house legal services or outside counsel in the matter. If working with internal attorneys, managers need to consider their past working relationship with others who will be involved in the project. If those relationships present problems, it may be beneficial to select an outside attorney. If internal counsel is selected, managers must clearly specify the attorney's role and establish appropriate communication protocols for all persons involved in the project so there is common understanding as to expected working relationships.

Selecting Outside Legal Services

Word of mouth still is one of the most common ways to identify counsel that is best suited to meet the utility's legal needs. Sources include utility personnel who have worked with counsel before, engineering consultants, other utility managers who have undertaken similar projects, regulatory agencies, and professional associations. Various legal directories that identify law firms by particular area of practice and by geographic location also may be helpful. Internet databases allow managers to search for attorneys by firm name, attorney name, clients, geographic locations, or type of law in which an attorney or firm specializes. Two Web site examples are www.attorneyfind.com and www.lawlinks.com. Many law firms also have their own Web pages.

Figure 12-1   A utility's attorney may be a manager's best asset

Requesting written statements of qualifications for legal services is helpful in selecting legal counsel and may be required in some circumstances. If the manager works for a municipal utility, the city attorney can provide guidelines and advice in selecting outside counsel and can cooperate in managing the selected individual or firm. A request should clearly identify the particular project or projects for which legal services are needed, without compromising confidential or sensitive information, and should clearly state any deadlines or time constraints for completing the work to be assigned. A request for qualifications may include some of the issues listed in Chapter 13, Support Services.

Managers should want to identify potential conflicts of interest when retaining outside counsel. There are two types of conflicts that arise with some frequency in representation of utilities: client conflicts and issue conflicts. Counsel should be fully informed of all facts that could pose potential conflicts as soon as possible.

The essence of protection against client conflicts of interest lies in the duty of loyalty to the client. A lawyer cannot represent opposing parties to the same litigation. However, counsel may ethically undertake representation with potential conflicts in other matters when certain criteria are met. Subject to certain prohibitions, counsel may undertake representation that involves a substantially related matter in which the client's interests are materially and directly adverse to the interests of another client or the attorney's or law firm's own interests if, and only if, the attorney reasonably believes that the representation of each client will not be materially affected and each potentially affected client gives prior written consent to such representation after full disclosure. When a potential conflict is identified related to litigation, the utility manager and prospective counsel will want to evaluate what dangers exist that could ultimately disqualify chosen counsel from representing the utility. Unfortunately, motions to disqualify counsel during litigation are a litigation weapon in some jurisdictions.

An issue conflict arises when counsel takes one position in its representation of one client and argues the other side of the specific issue for a different client. While there is no ethical constraint on issue conflicts, their potential existence should be discussed. In the selection process, managers must determine whether past representations would present an issue conflict and decide if such a conflict endangers the utility's position.

There are fees involved if the manager selects external counsel. Budgetary adjustments may be necessary if internal counsel is assigned. For any fee arrangement to be acceptable, it must be mutually beneficial to the utility and the attorney. Utility managers may prefer to have counsel identify exactly what services are compensated because of budgeting concerns. At the same time, the attorney must consider such factors as time constraints, complexity, degree of responsibility, likelihood the case will preclude other employment, fees customarily charged by other attorneys for similar work, and, where a contingent fee is involved, the likelihood of recovery. Managers should review applicable laws for any prohibitions on certain types of fee arrangements, such as paying or accepting a fee based on the passage or defeat of particular legislation or the outcome of administrative proceedings.

Fee arrangements are increasingly negotiable. The most prevalent billing formula for fee arrangement in utility and environmental practices is *billable rate × billable hours*. This is because issues that arise are often not amenable to a commodity approach where the lawyer offers a fixed-fee arrangement because the same product is sold to multiple clients. Billing methods that may be appropriate to particular utility and environmental matters include those described in Figure 12-2.

It is important for managers to prepare a detailed budget for legal services so that the utility and counsel can take a more strategic, cost–benefit approach to a

> ➤ Billable rate × billable hours. This can be different rates for each attorney in a law firm or a *blended rate*, which is an averaged rate for all firm attorneys

> ➤ Billable rate × billable hours plus *not to exceed* (to facilitate budgeting)

> ➤ Contingent fees (a percentage of the dollar judgment or settlement or a percentage of the administrative penalty saved; expenses and disbursements are billed directly to the client)

> ➤ Hourly rate plus contingent fee (negotiated either to lower the hourly fee or to limit the total amount of time to be compensated on an hourly basis)

> ➤ Premium hourly rates for designated activities (e.g., a higher rate for courtroom time)

> ➤ Retainer plus hourly rate (to encourage utility personnel to get preventive advice without the pressure of being billed by the hour)

> ➤ Flexible or volume rate (one hourly rate schedule up to a certain dollar amount and a lower rate for work performed above that amount)

Figure 12-2   Fee arrangements

matter. Budgets that are updated by status reports that track progress in the case and track expenses enable a utility to make informed decisions about its case. Status reports also reduce the potential for disagreements and misunderstandings over billings. Precise budgeting for any legal matter, particularly for those involving contested cases before a regulatory agency or litigation, is difficult. Contested cases are frequently unpredictable and subject to costs caused by the opposing party; events will occur that may force changes in a case plan or require emergency or remedial work and change the budget forecast. The key for this, and almost all counsel–client relations, is communication. Agree in advance to a procedure for communicating significant changes in a case plan or budget.

The chosen attorney's billing procedures, to the extent that they are compatible with the agreed-on fee arrangement, should provide the utility with sufficient information regarding the work performed by legal counsel. Invoices that disclose information described in Figure 12-3 encourage accountability. Specificity in billing may not be desirable in some situations. In some states, the details of invoices for legal services may be discoverable under a public information statute. Similarly, utility managers may not want personnel to find out about possible legal issues by reading an attorney's invoice. Managers should discuss any particular requirements in billing procedures before the first bill is sent.

Finally, utility managers need to have a written engagement letter or contract with legal counsel. Such a letter or contract should include the following elements:

- attorney in charge of the case

- potential conflicts

- fee structure and expenses

- retainer procedures

- specific billing procedures, if any

- procedures regarding timeliness of payments and any late fees

- procedure for fee increases

> ➢ Person performing the activity
>
> ➢ Detailed description of the activity performed
>
> ➢ Amount of time spent on the activity
>
> ➢ Dollar cost of the particular activity
>
> ➢ Total hours worked in the billing period for each person performing the activity
>
> ➢ Total dollar value for the work performed
>
> ➢ Itemization of any expenses charged

Figure 12-3   Billing disclosure requirements

- status report requirements, if any
- any privacy concerns

Contracts with legal counsel should accommodate the utility's standard contract requirements, to the extent appropriate.

MANAGING THE ATTORNEY–CLIENT RELATIONSHIP

It is important that managers understand who the client is. Generally, legal counsel employed or retained by an organization or governmental entity represents *the entity* as opposed to its board members or managers. In the ordinary course of working relationships, counsel may accept direction from a utility's authorized representatives but is expected to proceed in the best interests of the client. Counsel must honor his or her primary responsibility to the client if a potential conflict occurs. Individuals, such as board members or managers, should retain separate counsel in these situations.

The utility's attorney has certain ethical duties that must be performed. Lawyers are governed by standards of professional conduct. These rules address the client–lawyer relationship, fees, conflicts of interest, fairness in adjudicative proceedings, trial publicity, advertising, and many other aspects of professional practice. Managers can obtain a copy of the rules of professional responsibility for attorneys by contacting the state bar association or on the Internet. Most states or state bar associations also administer a network of staff and volunteers who handle grievances against attorneys for misconduct.

If a utility expects to receive effective legal advice, it is obligated to thoroughly inform its legal counsel regarding all circumstances related to the matter under representation. Attorney–client privilege allows open discussion and correspondence between attorney and client by protecting confidential communications from compelled disclosure to other parties in some instances. Attorney-client privilege specifically applies in instances where confidential communication exists between legal counsel and a client for the purpose of obtaining or providing legal advice.

Utility managers must remember that confidentiality must be carefully maintained. It may be waived or lost when the content of the communication is shared with third parties. For example, communications in the presence of regulatory agency personnel or engineering or environmental consultants are generally not afforded an expectation of privacy. Communications that are confidential between legal counsel and a client can be the subject of compelled disclosure if they are subsequently shared with third parties, whether intentionally or unintentionally. State public information or open records laws may have an impact on protecting attorney–client communications.

It is a helpful precaution to devise a records-management procedure with legal counsel in advance. For example, legal counsel may be asked to include a prominent warning against disclosure in any sensitive or confidential communications counsel generates. Accordingly, a utility may wish to segregate all communications to and from legal counsel within its files and e-mail records.

A good working relationship with legal counsel requires effective communication. Managers are responsible and accountable for conveying a real understanding of the utility's goals and measures of value, time constraints, and political climate to counsel. In turn, counsel should communicate a meaningful understanding of the prospects, costs, and uncertainties in the legal or regulatory process. Some of the items that should be discussed regularly with counsel to increase communications and project efficiency appear in Figure 12-4.

The need exists for periodic status reports to balance the difficulty in predicting long-term activities with a utility's need for budget information. Contracts with counsel should negotiate compensation for a reasonable time necessary to complete such reports. Managers may want to combine reporting with periodic meetings with counsel to discuss the status of a matter. A status report might include, but not be limited to

- name of matter reference or case name

- date matter commenced and anticipated completion date

- attorneys responsible for and assigned to the matter

- current case status, indicating activities and actions since the last report

- current posture of the matter and prospects for positive resolution of the matter or explanation of why the prospects cannot yet be evaluated

- updated budget projection

- any information required from the utility

A debriefing meeting with legal counsel is an essential part of the process. This allows the parties to discuss potential improvements in handling similar matters in the future, as well as an opportunity to review the results of the case.

SPECIALIZED MATTERS

Utilities often engage legal services to address special matters. In many instances, these require public disclosure of the engagement and specific information. Some examples include

- **Regulatory issues.** Counsel often joins utilities in appearances before regulatory agencies, both in routine and contested issues. These might include, but are not limited to, rate appeals, certificates for service areas, water rights, permit applications, claims, complaints about services provided. Such issues generally require that counsel have special knowledge of regulatory jurisdiction(s), rules, precedent rulings in similar cases, appeals processes, and methods of dispute resolution.

> Client's primary goals for the project and other acceptable outcomes

> Political and public perceptions of the matter

> Primary interested parties

> Budget and fiscal circumstances

> Primary contact persons and who should be copied on written materials

> Intended working relationship between outside counsel and in-house or general counsel

> Decisions that require advance approval

> Scheduling concerns, such as board or council meetings or budget preparation

> Client in-house capabilities that can save money

> Document production or control

> Special confidentiality or intellectual property concerns

Figure 12-4   Manager–counsel communication

- **Politics and legislative issues.** Many investor-owned utilities and municipalities maintain legislative liaison, either through lobbyists or their attorneys. Most governmental entities require such individuals to pay a fee and register as lobbyists to ensure public transparency. AWWA has recognized the importance of utility input into legislative and regulatory matters and has encouraged greater involvement of its members in providing water-related information and education through the political process. It is important that utility managers know the legislative issues of concern and that they communicate directly—and through legislative liaisons and lobbyists—their opinions on such matters. In general, public utility personnel may exercise their individual political rights but are precluded from doing so as an institution. Investor-owned utilities may establish political action groups that support candidates and fund legislative and lobbying efforts.

- **Litigation issues.** When a particular matter escalates to litigation, the utility may need *special counsel* to work with the utility's general counsel, providing out-of-the-ordinary expertise directly related to the matter under suit. The utility's general counsel may recommend special counsel, but the manager should ensure that the desired special knowledge is being retained.

- **Notifications.** Counsel reviews and ensures compliance with all public law for any notifications issued by the utility. This can include, but not be limited to, condemnation, mandated water quality announcements, procurements subject to legal notification, public hearings, etc. Routine notifications may require limited review, while more complex issues may directly involve counsel in the initial drafting of content.

- **Permitting.** Permit applications require counsel review of the application and representation before the regulatory agency during the application process. Counsel should have experience in appearing before regulatory bodies and have comprehensive knowledge of permitting requirements. The utility must ensure that the attorney is fully—and accurately—briefed on its capacity and on potential deficiencies that might impact the permit's approval.

The legal aspects associated with utility management continue to be complex and require inclusion of legal expertise at increasingly early points in management and operational issues. Whether legal assistance is a contracted service or a staff position and whether it addresses general issues or is specifically chosen in response to a particular situation, the utility manager must have confidence in the advice being offered and must closely monitor the costs and results of the service.

Chapter **13**

Support Services

OVERVIEW

No utility can retain sufficient staffing to address all components of the utility business. Utilities employ consultants and specialists from time to time for particular project development, for nonrecurring services, and for enhancement of staff capabilities. In this chapter, the focus is on

- internal support services and functions that are a routine part of utility operations and are normally performed by utility staff members

- external support services and functions that occur infrequently or that require special expertise not readily available through staff efforts

ENGINEERING SERVICES

Most utilities retain engineering services, either through a contract with a consulting firm or through their own staff. Engineering services include, but are not limited to, maintaining system maps showing all infrastructure; reviewing development plans and service capacity and completing such models as are necessary to determine major system projects; and inspecting construction projects. Internal engineering departments may provide design services for some projects or may supervise contracted professional services from consulting firms. Consulting firms may be selected to design, build, or manage the construction of major projects.

In selecting a consulting engineer, utility managers need to research the background of the firm and assess its expertise and experience in the work requested. Client references should be sought and checked. Proposed project staff workload and availability must be reviewed and analyzed, and the consulting firm must give reasonable assurances that it will assign staff to the project. These steps aid the utility in planning and sustaining the project schedule.

If the utility has its own engineering staff, the manager must clearly understand the impact the work will have on the utility's ongoing operations. It also is important to be sure the engineering staff is capable of performing the analyses, calculations, drawings, and specifications required by the project. Typically, some documentation

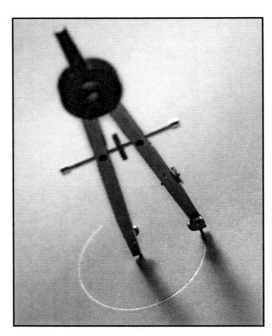

Figure 13-1   Engineering services are essential to a utility

(such as an engineering license or Engineer-in-Training certificate and documentation of additional training) to that effect must be available. The manager is required to provide information to the utility's governing body for decision-making purposes, so knowledge of the project parameters and calculations and how the engineer made decisions about issues is essential.

Managers must have working knowledge about four key areas in addressing engineering needs:

- **The big picture.** The utility's service area and projected service area govern the engineering requirements for the system. It is important that the utility develop, publicize, and implement service extension policies so that all requests for service go through the same review and approval process and that there is an established method for connecting to the utility's service system. In conjunction with the big picture and extension policies, the utility must develop and schedule a capital improvement program and determine the most appropriate financing mechanism.

- **Project planning.** This requires good research and projections concerning growth as well as knowledge of current infrastructure capacity and compliance issues. It also requires coordination with other agencies to address facility relocations when major street repairs are undertaken or when drainage issues are addressed. Finally, it requires that the utility manager conduct internal dialogue to be sure that capital programs are funded within the utility's rate structure.

- **Design.** Water treatment, water quality, and water distribution all are subject to advances in technology. As a result, utility managers must assess the best alternatives for their projects, keeping in mind the post-project requirements for operations and maintenance, staff capacity, security issues, etc. It is important that life-cycle costs are identified at the outset so that projects are promoted as cost-effective and able to meet the most demanding regulatory requirements. Public involvement in this phase often is a valuable tool—informed citizens are project allies and can lend credibility to the project.

- **Standards.** AWWA publishes a number of standards concerning system operations, materials, processes, and products. These form an industry- and world-wide accepted standard for use. More important, the utility itself must devote time and effort to determining the best practices it wishes to incorporate into its specifications. This sets the tone for quality components that are adopted and implemented as the approved business method for new development. Additionally, it is advisable to ensure that senior functional managers regularly review new products to maintain a cutting-edge posture.

PROCUREMENT SERVICES

Procurement requirements are generally spelled out either in corporate governance documents or by state or provincial and local government codes. Managers must recognize that those requirements differ and must be well-informed as to what they are and are not authorized to do in procurement. Most utilities have purchasing and contracting agents whose responsibility is to oversee annual and special contracts, receive proposals for professional services and for construction and materials bids, and process payments upon milestones within or at completion of projects or delivery of materials and supplies.

The primary responsibility of the manager is to ensure that laws and regulations are complied with and to establish the organization's procurement policies. These are often accompanied by codes of conduct or ethical requirements for dealing with vendors and potential contract recipients.

OTHER SERVICES

Many utilities rely on both staff and external vendors to supplement staff capabilities. They may retain firms for rate studies, compensation reviews, major construction projects, analytical services, etc. The utility must ensure that the firms they employ have the qualifications required to complete the contracted assignment and, most important, the utility must assign a staff member to manage the project for the utility. That requirement should include status reports to senior management no less frequently than biweekly, thus allowing monitoring of schedules, work progress, problems, and decision points at the earliest possible time.

KEYS TO SUCCESSFUL EXTERNAL WORKING RELATIONSHIPS

Regardless of the type of project the utility manager seeks external assistance for, there are two areas that are critical—selecting qualified firms and negotiating a contract that is responsive to project requirements and within budget. Any consultant's proposal should

- specify the names and qualifications of all persons who will work on the project

- describe their familiarity with the specifics of the project

- give examples of work projects that have involved similar or relevant components
- identify contact persons for each project
- review the firm's history of working in multidisciplined projects (such as finance, community involvement, multiple agencies)
- list potential conflicts of interest

Financial proposals are allowed in some circumstances, generally as a document separate from the proposal. In other instances, financial aspects of the project are negotiated after a firm is selected.

Chapter **14**

Gaining a Competitive Edge

OVERVIEW

Utilities and their leadership face continuing and increasingly complex challenges to

- make optimal use of resources and pay for much-needed improvements to aging infrastructure

- address expanded regulatory requirements by finding the best approach to increasingly stringent environmental regulations

- fund necessary programs in an environment that limits the expansion of revenue streams

- provide superior service that customers expect and control costs at the same time

- avoid the need for rate increases and provide competitive compensation to utility staff

A *best practice* is a methodology that, through experience and demonstrated use, has been proven to lead to desired results. AWWA's QualServ program has documented the common characteristics of better-performing utilities. In turn, when utility managers place emphasis on those elements, the result is increased internal and external confidence in decisions and actions. The most frequently identified characteristics from QualServ utilities are

- sound fiscal policies and asset management

- highly skilled staff and an investment in ongoing training and career planning

- an overriding focus on customer satisfaction

- a willingness to take risks

- a high credibility level in the community

Utility managers can get additional information about QualServ by visiting the QualServe section of the AWWA Web site, at www.awwa.org/science/qualserve/.

Commitment to using best practices must be utilitywide. There is a comfort level in "that's the way we've always done it" or TTWWADI. However, given resources available and customer expectations, everyone in the utility must have a knowledge base that supports the concept of best practices. Also, the utility must have available the tools, technology, and strategies needed to implement those best practices.

Utility managers in the twenty-first century must also focus on three strategies to gain a competitive edge. They must

- manage

- market

- merge

This chapter examines management practices, marketing opportunities, and options for merger.

MANAGING WITH BEST MANAGEMENT PRACTICES

A frequently heard opinion is that the private sector performs more efficiently and effectively because it is not subject to the bureaucracy of the public sector. However, although municipal organizations may have to comply with certain procurement and personnel requirements, all utility organizations—public or private—have bureaucracies they must deal with. A challenge for a competitive utility is how to manage the bureaucracy so as to continuously improve products and services. Many utility managers have begun assessing and applying techniques from private businesses in all types of situations. For example:

- benchmarking has moved from being a private sector/manufacturing terminology into water utility management. utility managers recognize the benefits of comparing their performance with that of their peers and, in some cases, competitors.

- strategic financial planning and asset management, a process that involves aligning financial strategies with other business process strategies, is an integral component of the planning process in most successful businesses.

- public and private partnerships are simply a public sector embodiment of the common private sector practice of outsourcing work that is outside the organization's core competency.

The use of these and other business tools by forward-thinking utility managers has allowed their organizations to become leaders in the industry and left them more prepared to meet their customers' needs.

In response to privatization and contract operations initiatives, many utilities are looking at themselves, how they accomplish their mission, what efficiencies are easily implemented that can rapidly reduce costs and improve work efforts, and what longer-term changes in business processes can sustain or enhance services. AWWA's QualServ, Partners for Safe Water, and other programs engage the utility in self-assessment with utility peer teams, resulting in suggestions for improved efficiency and effectiveness (see www.awwa.org). Vulnerability assessments offer another opportunity to identify and improve business processes. And, many utilities are required by their bond covenants to periodically engage an external consultant to conduct a management audit. Finally, each utility's annual audit provides an opportunity for the auditor to offer improvement recommendations through the management letter to the governance body. The governance body can instruct the utility's leadership to address the issues highlighted by the auditing firm.

Benchmarking and Optimization

Benchmarking and optimization are two interrelated processes that are often performed in sequence in an effort to improve the efficiency and effectiveness of an organization. Benchmarking is defined as a systematic process of searching for best practices, innovative ideas, and highly effective operating procedures that lead to superior performance and then adapting those practices, ideas, and procedures to improve the performance of one's own organization. A shortened definition may be best practices, and there are likely other definitions that are useful to water utilities in dealing with daily activities.

An outgrowth of benchmarking is the utility's ability to set goals for improvement after some measurement of a particular activity. These goals, or benchmarks, that the utility tries to achieve are determined or validated using other forms of benchmarking. For example:

- Thirty percent of a utility's customers inquiring by telephone about a bill are put on hold for up to 1 minute. This amount of time is determined to be unacceptable, and a new goal of 30 seconds of hold time for 10 percent or less of the callers is acceptable. Whether the decision is reasonable or not, this becomes a benchmark for future customer service activity. In this case, benchmarking describes a comparative analysis where the utility assesses its performance in one or more functional areas relative to another similar organization's performance in the same functional area. Managers should take care when performing this type of analysis to ensure that differences in the operational environments of the two organizations are taken into account.

- When comparing rates between two utilities, a discrepancy occurs if one is using the cash needs approach as the basis for rates and the other is using the utility approach. Comparison of the two rate structures without analysis of the methods of cost recovery can lead to erroneous conclusions. Municipal utilities may use different accounting and financial methods within either of the methods mentioned above. Managers must take care in drawing conclusions based on comparisons derived from different accounting methods if benchmark rates are to be cited.

Optimization, sometimes referred to as reengineering, involves changing the processes and practices within an organization in an effort to dramatically improve performance. Changed business processes can significantly benefit the utility's operations but require careful implementation in order to ensure employee acceptance. Benchmarking and optimization are often interrelated in that a benchmarking exercise is often performed prior to optimization. This identifies the areas that benefit the most from optimization and provides goals for the optimization process.

Strategic Financial Planning

Strategic planning (SP) is a tool that can help utilities meet the challenges identified at the outset of this chapter. SP is a process for assuring that financial and organizational policies and practices support the utility's strategic direction, that strategies are fiscally sound, and that they are appropriately implemented. Developing policies and strategies necessary to guide a utility into the future requires a great deal of organizational and individual time for preparation and interaction among functional units. Despite these resource requirements, there are significant benefits to the utility in completing SP.

Every utility is different and faces a different operating environment. However, all benefit from implementing a process that defines strategic direction for the utility and its business units, documents clear financial and operational policies, and assures consistent execution to achieve short- and long-term strategic and financial success. This is imperative in today's complex business environment.

First, the process enables the utility to assure sound policies and plans in accordance with laws, regulations, and good business practices. The emerging well-directed strategies let the utility meet stakeholder needs at an appropriate cost. Second, the management team understands and is committed to the financial and business plans and, therefore, makes consistent day-to-day decisions. This reduces the risk of inappropriate decision making. Finally, predictable organizational performance leads to better bond ratings, lower cost of debt, and predictable rates.

Strategic planning is a time-consuming and highly interactive process. The utility must follow these general steps in developing a strategic plan:

1. Conduct a utility analysis. Obtain a detailed understanding of the current financial and operational situation and all laws, regulations, and other considerations impacting financial and operational policies.

2. Conduct a senior-level workshop. Agree on the utility's vision, values, mission, overall goals, and high-level policies.

3. Draft detailed financial and operational policies. These should support the respective high-level policies and strategic directions.

4. Conduct a second senior-level workshop. This should be designed to develop utilitywide strategic initiatives and to review, revise, and finalize the more detailed policies.

5. Develop business unit plans. These should support the overall mission, goals, and strategic initiatives.

6. Develop a computerized financial model. The model should project financial results and the costs and benefits of implementation activities under different scenarios.

7. Develop implementation plans. These are the tactical steps assigned to each area within the organization that are to accomplish the utility's goals.

8. Finalize the financial, strategic, and business unit plans. A third senior-level workshop presents all information for review, adjustment, and agreement for implementation.

9. Implement financial, strategic, and business unit plans. The implementation phase places into practice the clear and accurate map laid out for the utility's future course.

MARKETING

Marketing has not been a core strategy for water utilities, regardless of whether they are publicly or privately operated. Marketing as a critical utility competency has emerged as the result of major national and international businesses moving into the bottled water market, the tendency for customers to want the convenience of water with them at all times and in all circumstances, and the increased emphasis on water quality. A combination of public relations and salesmanship, marketing is a support service that crosses functional boundaries and further expands the organization's outreach efforts.

Various stimuli trigger the need for aggressive marketing. Key customer needs are paramount. If a key customer requires a nearly fail-safe supply of potable water, it depends on the utility to ensure that availability. Marketing becomes a tool the utility can use to identify that need, determine the technical response to that need, and then closely work with the customer to meet that need (including the customer's role in achieving the objective). If a utility experiences a water quality problem, affected customers expect that they will have sufficient potable water throughout the duration of the incident. Marketing becomes a tool to identify how the water will be provided—whether through utility-bottled water capacity, tanker deliveries, agreements with private-sector vendors, or other alternatives.

As noted in chapter 3, marketing is not a reactive strategy. Marketing the utility's clean water and service is an ongoing effort that must be publicized as part of the strategic nature of the business plan. Just as any organization—retail merchant, school, church, grocer, service company—must carve out a niche, so must the water utility industry, especially in a time of increasing competition and changing loyalties. Traditional marketing strategies, including pricing, purity, reliability, do not resonate with customers accustomed to convenience, high-profile spokesmen, and perceived higher quality.

Marketing does not require a high-dollar expenditure, but it does require creativity and a willingness to innovate. Staying current with trends in the utility's service area and being well-positioned compared to other utilities is a commonsense best practice. Answering several questions can help managers enhance the utility's image.

- How does the utility perceive itself?

- What role does the utility play in the community?

- Are the utility's mission and vision statements current?

- How and where does the utility communicate what it is, what its role is, and what its mission and vision statements are? Answers to the last question may include the utility letterhead, bills, vehicle and equipment stock, business cards, public information releases, and conservation materials.

With credible answers to these questions, managers can devote the necessary time and resources to expanding the utility's outreach efforts, improving public perceptions, and focusing on those actions and activities that bring the utility the greatest benefit. Other important questions relevant to the utility's marketing plan include the following:

- Who are the utility's critical customers?

- Do utility representatives visit critical customers on a regular basis?

- Does the utility have a "Critical Care Team" customers can contact directly with questions, needs, and emergencies?

- Does the utility recognize critical customers in any way?

- Who are the utility's critical employees?

- Do managers visit them regularly?

- Are they recognized in any way?

- Does each employee have business cards?

- Is there a mechanism that allows employees to suggest improvements?

When integrating customer needs and expectations into the utility's marketing effort, it is important that managers focus on the majority of customers, not on the vocal minority. Managers must stick to the utility's vision, mission, goals, and objectives and carry them out in an effective and efficient manner—and make sure these efforts are widely known through internal and external communications programs.

DEVELOPING ORGANIZATIONAL ALTERNATIVES _____

One of the best-practice options available to utility managers is to explore merging or establishing joint action agencies. This can be a particularly effective approach where adjacent smaller systems (municipal, nonprofit, etc.) need to expand their service capacity (such as by developing additional water resources and/or constructing a water treatment plant). While there can be resistance to collaboration based on control and territorial issues, joint action agencies and strategies can enable utilities to

- join in procuring commonly used materials and appurtenances in order to achieve economies of scale normally reserved for larger utilities

- join in offering a common service center for billing, inventory management, and support services, such as dispatch, safety, equipment repair, and human resource support

- join in construction of major infrastructure, such as treatment plants and storage facilities

In those instances where proximity and organizational culture are similar, merge two or more separate entities into either a single utility or into a jointly governed agency that provides some or all of the services required by some or all of its members.

In this environment, many utilities are pursuing alternative delivery options as a way to incorporate industry advances in improved services and to reduce costs. Traditionally, utility employees have operated and maintained municipally owned or privately owned facilities that are constructed using a design–bid–build procurement approach and that are financed using proceeds from tax-exempt financing, rate revenues, grant funding, or appropriate types of conventional loans and bonds.

Alternative delivery systems have entered the utility business model, increasing private sector participation in the operation, management, and, sometimes, ownership of utilities. While private sector companies that specialize in providing water and wastewater service may be able to achieve efficiencies and cost-effectiveness, there is also evidence that public utilities that undertake quality improvement

Figure 14-1   Managers must be sure to select the right partner

programs are able to match those benchmarks. Frequently, those efficiencies are achieved by outsourcing noncore functions, such as custodial or landscape maintenance services, equipment maintenance, and meter reading.

Another alternative delivery approach is contract operations, in which the utility enters into a contract with a private partner for the operation and maintenance of one or more components of the owner's utility system. Such contracts have terms ranging from 1 to 20 years. It is incumbent on the utility to fully understand any contractual requirements and to ensure that its interests are fully protected should higher costs emerge as the contract period progresses.

A third approach involves the design–build concept in which the owner hires a design–build contractor who is responsible for both the design and construction of a new facility and who accepts full responsibility for any cost overruns during construction. A variation is the design–build–operate concept, which expands the process so that the contractor operates and maintains the new facility under a long-term operating contract. The contractor has an incentive to emphasize operating efficiency during the design and construction phases, and the owner can reasonably expect lower life-cycle costs for the facility. Another variation includes two additions—ownership and financing. In this model, the contractor finances the project using debt and private equity and recovers construction costs as part of the fees paid by the utility over the term of the operating contract.

The Pros and Cons of Alternative Delivery Systems

Significant changes in the ways utilities finance and deliver projects have both advantages and disadvantages. Advantages include operation efficiencies and reduced operating costs, shared risks or risk reduction (including environmental, force majeure, operational, and financial), lessened politicization of utility issues as a result of contractual obligations for necessary resources, reduced construction costs that result from collaborative efforts of the owner and operator during design and construction phases, and access to private financing or capital not normally available to public sector utilities.

Disadvantages that can adversely impact utility management include a possible decline in service quality, potential loss of control over the assets used to provide an essential service, higher costs for financing, and deterioration of assets unless contracts require an appropriate investment in the system throughout the life of the contract. In addition, the utility must consider whether the introduction of privatization to a workgroup within a larger functional area will significantly affect the operation of the overall department.

Managers may consider alternative delivery systems for their utility when exploring all of the options that are available rather than continuing to do the same old thing that gets the same old results. However, managers should remember that the utility is always the focal point if there is a decline in the quality of products and services, regardless of whether the utility itself or a private entity is in charge of operations.

The utility's current business model may be appropriate and only requires slight modification. However, a major change in the business model may help the utility operationally, administratively, and financially. Indeed, managers should step back and objectively review the business model. They may find that the utility does not have one, that employees and customers do not know what it is, or that steps were never taken to evaluate how business operations and practices measure up against the utility's business plan.

The bottom line is that managers do not know the best alternative for their organization unless they explore all available alternatives. In general, utilities look to alternative service delivery systems as a way to

- address recurring operational issues

- gain expertise they do not have internally

- take advantage of cost savings that private partners offer

- gain access to private financing options

- reduce the politicization of decision making

Alternative delivery systems can be an effective means of addressing some of the challenges that water and wastewater utilities face, but they are not always the right answer. Many practices used in the private sector are transferable to public utilities, who must fully examine the possibilities offered by reengineering and optimization before pursuing an alternative delivery system. The best advice from those who have engaged in the process is direct and meaningful—*select the right partner*.

Index